The Cost of the Car

Human, Environmental, and Economic

Ian East

Premier Edition

Open Channel Publishing

www.openchannelpublishing.com

The Cost of the Car : Human, Environmental, and Economic

A *View from the Lab* published by Open Channel Publishing Ltd. in 2010.

ISBN: 978-0-9565409-0-4 (premier edition, paperback, monochrome).

ISBN: 978-0-9565409-1-1 (premier edition, hardback, monochrome).

ISBN: 978-0-9565409-2-8 (premier edition, e-book, colour).

www.costofthecar.info

www.openchannelpublishing.com

To my father, Deryck East (1921 – 1989)

for the opportunities you helped me find
and never found yourself

To my mother, Patricia East (née Fuller, 1925 – 2007)

for showing me just how wrong the establishment can be

At the top of each their own mountain are found two cities, named Heaven and Hell. Despite the chasm between, a citizen of one can easily pass to the other, across a gossamer bridge.

All those in Heaven know they are in Heaven.

All those in Hell believe they are in Heaven.

Traditional

Content

About the author

Ian East began his career as an Engineering Physicist in the defence industry before returning to university to gain a Ph.D. in physics from Imperial College, London, in 1983. He then continued as a University Lecturer in physics, and later computing, in diverse institutions over the next twenty-five years, publishing two textbooks and around thirty papers. He has also served as a consultant and contractor to commercial and military institutions.

He now spends his time writing, publishing, home educating, and implementing his own programming language, Honeysuckle.

He lives with his wife, two sons, and a large Labrador, near Oxford.

Preface

In the late summer of 1991 I found myself sitting on the edge of the bed in a San José hotel, wondering why they fitted letter-boxes at all to a hotel-room door, never mind such big ones. As I got up and turned towards my computer, a monumental thud announced an answer.

A typical British local paper comprises a couple of pages of parochial news, followed by a few more that advertise jobs, of the low-paid variety — not the most riveting stuff, as a rule. I think I stared for a full five minutes at my first experience of an American version. Now I understood the jokes about people being decapitated when these things are thrown.

Once in a while, you read something that causes a tectonic shift in your understanding of the world around you. To this day, the plates are still colliding, and mountains still rising.

At the time, I no more questioned the value of the car and its place in our lives than I did that of tables and chairs. It was how you got about. You *had* to have one. This did not even get as far as needing an assumption. My little GTi was almost part of me, and I took tremendous pride in being able to exploit its ability to corner tightly at speed. I would even use the gully on the inside edge of a bend in the road to get around a little faster. I had never given the slightest thought to the possibility of being denied this liberty.

On the front page stood a woman looking a tad dischuffed. One minute, she had been a normal single mum laying into the washing, with her eye on the clock for the school pick-up; the next, she was on the floor looking up into the face of a worried paramedic neighbour. It was a very mild form of epilepsy that had matured. She might expect similar blackouts perhaps just once a year. But that was more than enough to lose her licence.

To use her own words, she had become more of a cripple than if she had been confined to a wheelchair, whereupon she might still be free to drive. Instantly, she had been deprived of the ability to get to work, to collect her children from school, to do her shopping, or to reach *any* amenity.

Moving was not an option that offered any solution. The entire area had been developed assuming universal access to a private car. As a result, no public transport system could be truly effective, or viable without heavy subsidy. Any worthwhile service would entail a large fleet of vehicles, each slowed by frequent detours, and carrying very few passengers. Her nearest stop was more than a mile away, and buses there would follow only a single route. It would take hours just to get to the 'local' supermarket.

Provision for cycling was non-existent. In any case, the distances involved would severely challenge anyone out of condition, especially when laden with a child or shopping.

I slowly began to wonder about the lifestyle of all the others with minor but disqualifying infirmities who, prior to the car, would have been indistinguishable, leading full and normal lives. I thought also of the young, who are denied any independent mobility until full grown.

I began to ponder what they lose along the way.

In such places, non-drivers are completely disenfranchised. The right to own and use a car has been celebrated to such a degree that the right not to has been utterly destroyed. And such places are everywhere now.

Hardly a week goes by without news headlines that reflect one or other of the issues raised by the dominance of the private car over our lives — damage to the environment, congestion, or the hideous carnage it causes, for example. Given its prominence in the media, there are surprisingly few books on the subject. Those available are either academic (concerning general transport policy, economics, or planning) and somewhat tangential to the car itself, or are hostile to the car on principle.

My intention here is to present a *balanced* view of the subject for those who, like most of us, have always accepted the car, but now harbour doubts. Anyone seeking a rant against metal monsters may be disappointed. While they will find much ammunition, they will find no call for the State to terminate our liberty to own and drive our very own personal chariot.

On the other hand, I see no need to sing out the *benefit* of motoring. I could not hope, in any case, to compete with a professional advertising industry, well funded by car manufacturers, oil companies, and a host of other traders. All its advantages are acknowledged and accepted here, without undue criticism. It remains only to dispel certain myths: for example, that motoring is cheap, or that it is open to everyone.

What moved me to write this book was a series of quite unpredictable events. It began with encountering the story of that woman in San José. A year or so later, I made a move to self-employment, and found myself better served by bike and train. I sold the car and began four years without one. While I missed driving, my general health and fitness improved, my savings grew rapidly, and my life became more relaxed, social, and productive.

It all came as a great surprise.

I began to read widely, and think deeply, about this modern 'necessity'. As a scientist, I began by assessing it *quantitatively*, in comparison with other forms of transport. The extent of its inferiority does not take long to discover, even when expectations are limited to speed, safety, and economy. As time passed, ever more concerns came to light, and its full impact on our health, community, and environment, became apparent. The dominance of road transport even appears as a block to economic growth because it is inherently incapable of automation (despite what some still claim).

What I had expected to be a relatively short book, simply comparing modes of transport, grew to take over half my life for nearly six years.

I have resisted delving too deeply into the politics of motoring. Its sociology and psychology I also leave to others. My concern is the practical.

Lastly, I wish to avoid adding to the existing pile of academic works on the subject. On the other hand, most of those are quantitative only in that they quote statistics. I found only one author who gave any formal (mathematical) analysis.[1] A true and proper account of anything that varies — and traffic certainly does vary — requires this. While I have spared the reader all the gory details, I have tried to lay open the nature of the important relationships involved and all the significant consequences.

I hope to provide the first popular, objective, and (more or less) complete, account of the true cost of our almost total reliance upon the private car.

1 I must thank Martin Mogridge for his seminal insight, for performing all the analysis we need here, and for saving London. (See Chapter 2, Section 3, and Further Reading.)

Chapter 1 begins the book with a reflection on family life in the early 1960s, as the car was becoming ubiquitous, and railways were being closed.

Chapter 2 starts by addressing just what we want out of transport, before going on to compare the various means available for the kinds of journey we make most often. It then looks at why and how congestion arises, and what can be done about it, with regard to market pricing and competition.

Some might find this last section a little challenging. Though they may prefer to skip over the explanations, they might nonetheless note the conclusions. These form principles of transport provision, and are able to explain many of the results of transport policy over the last sixty years.

Chapter 3 addresses the most obvious and appalling cost of the car — the daily carnage it causes. Statistics fail to reflect the true price paid, or the threat posed to each and every one of us, and our loved ones. Attempts to quantify the loss and misery in purely economic terms are similarly inadequate. To unearth some semblance of the truth, three short stories are presented. They portray genuine road-traffic accidents, but with fictional characters replacing the true victims, to protect their identity.

Chapter 4 deals with the energy crisis and the environment, both in general terms and with regard to the car. In particular, consideration is given to the consequences of heavy reliance upon cheap foreign oil, the effect of carbon emission, and various forms of local pollution around roads.

Chapter 5 starts with a summary of all that has gone before. Recognizing the inseparability of transport and the pattern of land use, various examples of urban planning are then described, from the Garden City of Ebenezer Howard, through Le Corbusier's Villes Contemporaine et Radieuse, to Walt Disney's Experimental Prototype Community of Tomorrow (EPCOT).

A profound need for quantitative, as well as qualitative, design emerges. Though the city was never built (only a theme park), Walt Disney is honoured, both for his recognition of the problem and for his valiant and visionary fusion of art and engineering in addressing it.

Finally, consideration is given to how our habitat might be improved, with regard to both new and existing structure. A summary is given of the New Urban movement, as an example of a new approach that responds to many of the problems posed by car dominance. A slightly broader set of principles is proposed, to both reduce demand for mechanized transport and to meet remaining demand with a minimum overall cost.

Some of these ideas might seem somewhat removed from the issue of the private car. But the simple fact is that the car is not just the *cause* of much that ails us: it is also both symptom and effect.

1

Progress

1.1 Home and away

I remember an English spring sun shining as I skipped down our driveway. My school day started with a ten-minute walk, on my own, after Mum had helped me across the road.

My little journey began with the pub, which sat stolidly opposite. Third down was the shop where my elder brother worked two evenings a week "stocking up" in return for ten bob and an occasional tin of Ambrosia creamed rice, devoured in a single sitting. Next was the timber yard — scene in a later year of a great battle between blazing wood and brave local firemen. Then followed a long row of varied bungalows and cottages, each competing to produce the loveliest frontage. Low walls topped with iron railings, white picket fences, and roses, set or softened the scene. I would then pass the library — by then a shop, but still known as The Library — and a butcher's. A bakery marked the final turn into School Road.

Of course, all of this was taken utterly for granted by a six-year-old. It had all been there since the dawn of time, like the hill opposite.

Now and then, Mum would meet me after school and we would walk another five minutes down to the village green for some shopping. Here, there was a larger butcher's, another bakery, a grocer's, pharmacy, electrical store, post office, two barbers, three more pubs, a fish-and-chip shop, and a small church, where Mum's mum and dad were married.

There was also Jasper's.

Jasper was a small gaunt old boy of an age anything between fifty and eighty. Again he must have been there since grass began growing. There was no outward sign of trade. You had to know how it was done. Finding him was no simple matter; he had the ability to appear and disappear at will, like a ghost. You need then convey which bit of what machine had let you down. Silently, he would then depart into a dark and cavernous recess, amidst what to anyone else was a gigantic mound of scrap. Without fail, he'd always return with something that would do the job.

We would meet neighbours and friends, out and about. Somehow, days were longer then. There was always time for talk.

Figure 1.1 The village *c.* 1955 (©The Francis Frith Collection).

My most treasured memory is the Friday afternoon visit to Gran's, and the house in which Mum grew up. We would set off down the lane at about three, and turn across common land towards their village, just past one of the paper mills that punctuated the stream. The grass was taller than me, so all I had for navigation was Mum's hand. Beyond the fields, we would cross the main road, and climb the embankment on the other side. This was topped by the 'new' railway, on which wonderful things could sometimes be seen. A steam express would occasionally thunder past, pulled by a Class A4 — perhaps even the record-holding 126mph Mallard itself.

Shortly after we arrived, my dad, grandad, aunt, and uncle, would come home from work and join us for tea. Family lay within walking distance, and work within the range of a bicycle.

At seven, I moved up from 'Infants' to Primary School, in the next village along. Money was tight, and I was to have no bike until I was ten, so I walked (or ran). By this time, I was trusted to cross the road alone. Traffic was still light on a country lane.

It was a Church of England school, built from flint in an old shire style, matching the massive church standing opposite. At least a century old, it had no ceiling and a high roof that leaked when it rained. But the headmaster was second to none. On a balmy summer day, he would sometimes interrupt lessons and take us across the road and up into the meadows, and tell us all about the flowers and insects we encountered.

There was no misbehaviour. It simply couldn't happen. "Zero tolerance" went with a charisma and an automatic command of respect. He knew each of us personally, and would visit our homes, out of the blue, to talk to our parents. We were unaware of the absolute discipline. All we knew was the warmth of his smile and the same sense of security we had at home.

The simple joy he took in his job spread easily to his staff.

I made several friends, and would occasionally walk to their homes to play. Many an errand did I run at the weekend, to exchange garden produce with neighbours, or to the shops for something missing in the larder.

No one questioned the safety of a small child out alone.

Age ten brought me to secondary education. I had passed the 'eleven-plus' exam a year early, and thus gained a place at a Grammar School in a small town about eight miles away. There was never any doubt that a child would get their self to a secondary school, independently. Anyway, I had been travelling solo since I was six.

Now, this meant using the Great Western Railway.

I had grown up with the steam whistle, and the regular mechanical clanking of steam engines. "God's Wonderful Railway", as it was known with great affection, ran past our back garden. Though this part had always been single-track, it was no branch line. It linked the two largest towns in the area, and sported its own branch alongside the Thames to my new school. To use it on my own made me feel quite grown-up.

The little station above The Green was an absolute delight. Like many railwaymen, our stationmaster took great pride in his gardening, and had won awards for the floral display facing the platform. Like stationmasters everywhere, he held a respected position in the community, and returned this with a smile and a welcome. In summer, there were his flowers, and in winter, a friendly waiting room, warmed by a well-fuelled stove. One day, as I enjoyed the flowers, a little girl, some years my junior, sat down beside me and calmly announced she would marry me when she grew up.

I often wonder if she did better, in the end.

A railway station was then more than a place for personal transport. There was a goods yard. Everything from parcel post to a wagon load of coal would arrive there. It was the lifeblood of many a local business.

Several schoolmates joined me along the way. From an all-boys school, we came to welcome the presence of high-school girls at the junction. They got on where we got off, alas, but *something* was learned, nonetheless.

The old tank engines were soon retired and replaced by characterless diesel railcars. One advantage of these, however, was easier communication with the driver *en route*. In the absence of a guard, they could sometimes be persuaded to stop and let me out at the back of my house. If they changed their mind, as they sometimes did, I'd have a twenty-minute walk back from the next station along.

There was a lesson in that as well.

My life changed entirely with the arrival of my first bike. Suddenly, I was free to go wherever I wanted, whenever I wanted. Like most lads of the day, I thought little of riding five miles, and occasionally rode ten. A friend and I used to ride simply to explore the countryside, and for the fun of it.

It became even easier to get to The Field, where local lads would meet for football or cricket, and older ones pass skills on to the younger. There were several places we could play, but the playing field of my old Infants' School was the nearest, within ten minutes' walk, or two minutes' ride, of home. We would be there till dusk, the year round. Our parents felt no concern over our safety or behaviour. There was never any trouble.

We were happy, and not one of us was the slightest bit overweight.

It is perhaps easy to be a little starry-eyed about your own childhood. As a child, you were ignorant of all change and all that could be changed. You were sheltered and protected from all the many things that could go wrong.

Once in a while, reality threatened to break through. I remember one of my cycling forays led me to explore a wood, a long way off. We found something buried in the brush. It turned out to be the cockpit of a World War II bomber, more or less in tact, but with clear evidence of fire and destruction around it. It never occurred to me then that it might well still harbour human remains. Dad had served as an electronics technician in just such a machine. His brother had been aircrew, shot down over Germany, and imprisoned for years. I made the connection rather suddenly.

Much earlier, there was the evening at Gran's when the Cuban Missile Crisis was reported. I was just six and had no understanding at all. But I had a child's sensitivity to their guardians' sense of security, especially when it collapsed. My grandad, who had ridden despatch between the trenches, and watched doodlebugs descend, merely muttered "here we go again".

His tone and expression said rather more.

Despite this and the many other 'challenges' of the era, it was perhaps the best time and place to be alive in all history and all the world. There was neither famine nor pestilence, and peace prevailed. There was housing, a health service, and education, for everyone. Even an urchin like me could eventually go to university. All the benefits of mechanized industry were finally filtering down to ordinary people, without any of the environmental consequences yet appearing. There were Walt Disney movies, and wonderful music. I will never forget the magic in hearing each new Beatles and Stones '45' on my elder brother's Dansette, and later, Marvin Gaye singing *I Heard it Through the Grapevine* on the radio, in the back of Dad's Skoda.

Harold Macmillan was a only a little premature, in 1957, when he told us we had "never had it so good". A TV and a fridge were soon to come.

And every man could soon possess his very own private car.

1.2 All change!

The little shop, where I and my brothers worked, survives, though in different hands. It still incorporates a sub-post office, and is of tremendous value to local people, especially the old and infirm. The timber yard has fared similarly well. Arguably, the greatest change is visual. Parked cars now dominate the view. Gone are the front gardens. They have *all* been replaced with car parks, in harsh brick, concrete, tarmac, or shingle.

Once a picture, the lane has become ugly.

The aural experience has also changed. Noise from the now busy road mounts a tireless assault. Life is no longer tranquil.

It now has a breathless pace. Time seems in short supply.

There are fewer opportunities to meet, fewer places to walk.

FIGURE 1.2 The village *c.* 2005.

The nearby butcher's and The Library are now just houses. The Green retains some of its amenities, though the butcher's, grocer's, bakery, electrical store, post office, and both barbers, are gone. Their place has been taken by a supermarket outpost.

Nobody could replace Jasper. Nobody tried.

Mum spent her last few years with both hips replaced, but enjoyed life to the end, especially café culture. Not without difficulty, however.

Her country lane was reclassified a trunk road, although it was never widened, and became choked by parked cars. She was forced to cross it, even to reach the bus-stop on her own side, because many of her neighbours park on the pavement, often blocking it completely. Though I twice complained, the police took no effective action.

Many a home now boasts two cars. Rarely are both accommodated in driveway or garage. It seems quietly accepted that we can now treat not just the road outside our house as our own, but the pavement as well.

Provided we get there first, of course.

Parking on the pavement is not specifically illegal in Britain, but driving on it certainly is. If the services of a Boeing Chinook military helicopter, or similar, can be dismissed, parking does indeed directly infer an offence.

FIGURE 1.3 The pavement becomes a car park.

The police, it seems, now decide which laws to enforce, and which not to. This has precisely the same effect as deciding the law itself.

Having never learned to drive, Mum became dependent on buses. With only a minority to carry, these are reduced in number and depend heavily on subsidy. They are also much slower, because of heavy congestion in and around all local towns, and because they stop more often to serve the continuous "ribbon development" along the way. Fewer people are collected at more frequent stops. Despite the subsidy, tickets remain expensive.

Neighbourliness declined and antisocial behaviour arrived, though property prices soared. Mum learned to lock her doors and windows.

Her brother and all three of her sons moved away.

Every last piece of land has now been 'developed', including the common across which we walked to Gran's, and the school playing field on which I had so much healthy outdoor fun. There is a park remaining, but too far away to use much. Yet there are far more children.

It is no longer safe for anyone, least of all a child, to ride a bike.

The truth is that it was already unsafe to ride a bike, or to cross the road, forty years ago. There was just a different attitude to accidents and safety back when much of industry was still fraught with danger, and war a vivid living memory.

To get home after Infants' School, I would merely retrace my steps. Mum would wait at the foot of our drive to see me across the road. One day, I misread her signal and stepped out. A screech of brakes announced the baker's van, as it managed to stop within a foot of running me down. I remember only a strange silence as I watched both Mum and the baker pulling their hair and yelling.

One of my classmates was not so lucky.

She was not even *on* the road when it happened.

It was typical of many such accidents. An uncontrolled dog ran into the road. A driver swerved to avoid it and mounted the pavement.

She was holding her mum's hand at the time.

I remember the empty desk as our class was told.

An elderly neighbour of ours lost one of her children to the road, at about the same time, along the path our children walked to school.

Roughly, two children each day still die on the road, in Britain.

Twenty a day are seriously injured. Brain damage is most common.

We tend to think of road accidents as affecting only those within a car, but a quarter of the casualties are innocent bystanders — cyclists and pedestrians — all too many of them children.

Soon after I lost my first classmate,[1] my brother, aged eleven, saw a child crushed to death beneath a three-ton truck just across the street.

While the vast majority of ordinary folk managed happily without a car in the early 1960s, there were already very many on the road. From 1960 to 1967, in Britain, though the number of trucks remained steady at around four million, the number of cars doubled from five to ten million. Why the road casualty rate did not rise in proportion remains a mystery. One possible explanation is that people simply withdrew, out of the way. They simply walked less, cycled less, and became wary. Life moved indoors.

Childhood has quietly changed beyond recognition.

Besides the dramatic increase in the number of people, houses, and cars, there has been one other major change. The railway, that had laid outside my bedroom window, is gone. It closed in 1970 after serving many a town and village for more than a century.

With the railway gone, the car became a part of my daily life. Getting to school by bus was possible but hideously slow and expensive. My bike might have been an option, but by then the extent of the risk posed by motorized traffic was becoming clear. Having travelled to school independently from the age of six, I was to become dependent on adults at fourteen. It was the birth of the "school run", that now accounts for a quarter of all car trips in the peak morning period.

Provincial railways could neither suffer the loss in custom, that came with the rise in private motoring and commercial road haulage, nor adapt to serve the growing 'sprawl'. Nor could a single-track line, that would meander gracefully around every hill, compete with the fast new duplex one to the city. Like the Romans, the last of the railway builders drew straight lines on the map.

At the time, no provincial town could afford to remain disconnected from the network. Lines were built even if their economics, in isolation,

1 Two more school friends were to die on the road while I was in my teens.

were doubtful. Larger companies would later find it much cheaper to buy them up, as they went broke, than to extend their own network. The Great Western Railway was no exception, and found itself with laborious paths that strangely complimented Brunel's direct route from London to Bristol, earning it the nickname the "Great Way Round".

Our line remained open a little longer than most in order to allow the construction of a new motorway bridge across the valley. The steel girders were too heavy to be transported by road — even along the motorway itself. They could only be moved by rail.

Their eventual replacement will cause a bit of a problem.

This bridge has all the architectural appeal of a giant urinal, and is built over the once lovely meadow I crossed on my blissful Friday-afternoon walks to Gran's. Promises to deny development alongside proved worthless. Huge bright-red warehouse doors now adorn green wooded hillsides. The air has been turned from fresh to foul. Noise is pervasive and incessant.

Curiously, the motorway has repeated what the railway once did, though without the elegance or efficiency. People have arrived in large numbers, driven by the desire to live somewhere more pleasant, while retaining their well-paid jobs in the town and city. Only, instead of escaping urbanity, they have brought it with them, while adding a typical thirty-mile commute.

The Thames Valley has been prosperous for a very long time, for reasons rooted in transport. Before road and railway, there was the river.

Commerce arose in South-Eastern England around the Thames, the mouth of which provided an excellent port for international trade. Goods could always move easily along a river, which is long and has many tributaries. I grew up in an area that began rich in natural resources. Forests provided timber, for trade, construction, furniture making, and pulp. Paper mills employed many here for centuries. A river provided not just transport, but a sustainable power supply for industry as well.

Only the railway offered more.

Even as a child, I could not be convinced that this ugly, noisy, motorway was any worthy replacement. It is a simple matter to show, as we later shall, that no road can compete with the speed, capacity, or value for money, of a railway. To explain why motorways were built, and the railways allowed to decline and close, is harder. It was never a purely commercial matter.

The predominant reasons were political, not technical or economic.

I was very lucky to enjoy the regular company of grandparents, uncles, and aunts. They were a big part of my life, and a major source of security. They were also able to support each other in ways that go back to the dawn of time. All this has now been condemned to the pages of history.

My children cannot enjoy their extended family in the same way, nor can we draw the support our parents knew. One of my brothers lives a hundred miles north-east, the other a hundred south-west. We meet at funerals and the odd wedding. The cost involved usually dictates that I go alone. Christmas gatherings have all but ceased. The only surviving grandparents live forty miles away, and my only 'in-laws', with two children of a similar age to mine, are now two hundred miles south-west.

We are becoming a family of strangers.

My brothers and I have just endured the inevitable task of burying our last surviving parent. Our mother was laid to rest in the same cemetery as her husband, both her parents, at least one uncle, and no doubt several friends and more distant relations. An aunt and two grandparents lie in the cemetery of a neighbouring village. All traceable family roots can be found within ten miles or so. We will rest elsewhere, and far apart.

Something profound has happened to the way people relate to place.

Perhaps my brothers and I felt less bound to the area than did our parents and grandparents, though I shared a deep affinity for it. The invasion by an affluent effluent from London pained me deeply. It changed the character of town and village, making both insufferably busy. Within a decade or so, an "after you" culture had been replaced by one of "after me".

Nowhere was this more evident than on the road.

Above all else, the new influx, probably like the last, that arrived by rail, caused the cost of housing to escalate utterly beyond the reach of most local people. Ultimately, my brothers and I were forced to move away, in order to afford a roof over our prospective families.

New transport infrastructure, particularly that which follows the arrival of a new mode, has always caused a migration, of people, ideas, and capital. It is happening again now across Europe after the introduction of cheap flights. Brits are "snapping up bargains" all over Europe, particularly in the East. Equally, labour is on the move in the other direction. Property prices rise in the East and pay falls in the West.

Such migration began with the industrial revolution. A digital version of the British census of 1841 has recently appeared on the internet, making it easy for us to trace our ancestors. By far its greatest appeal comes with the recognition that the location of your family then is almost certainly where they lived for the previous thousand years or more.

To some, leaving meant liberation; to many others, it was exile.

It is all too easy to measure what was *eventually* gained; what was lost, in the association between people and place, we shall never really know.

The village in which I grew up suffered virtually no crime or antisocial behaviour. We never even locked our back door, still less our windows. It is a very different story in the mire that my old home has become. By the 1980s, I found myself watching drug deals undertaken in the pub car park opposite. More recently, the local shopkeeper has been robbed at gunpoint. At the village green, the proprietor of a shop was beaten during a terrifying armed robbery. There have been two recent murders.

Not a place in which I would choose to raise children.

I have been lucky enough to find and afford a home in another village, further from the city, in which to do so. While, like my old home, it knows no indigenous crime or antisocial behaviour, such problems leech in from nearby sprawl. A neighbour thumped on the door late one evening, after we had tucked our two up for the night, to warn us of a car on fire outside our house. A fireman later casually informed us that the creatures responsible often 'torch' a vehicle they have stolen, just to destroy finger prints.

A charred teddy bear lay lost among the wreckage.

There were three other stolen cars parked out front that night, prompting another neighbour to quip "it's getting hard to park around here".

The police lacked sufficient interest even to show up.

My gran was born in the same village as me, while Queen Victoria still ruled the greatest empire in history. Her life began when people got around on foot, bike, or train, and only horses powered vehicles on the public road. She saw the first cars arrive, driven by men so rich they paid income tax.

Railways were still being built, including the 'new' 1910 short cut from London to Banbury, which gave Great Western a faster route from London to

Birmingham. Gran would have watched as an army of 'navvies' descended, toiled, and left. A few of them died there, when a tunnel collapsed.

A little while after they married, and my mother arrived, they bought a new home halfway up the hill above the old London Road and below the new railway. This lovely spacious modern home gave a grandstand view of the valley below. From here she witnessed the first airliners fly by. As propellers gave way to jet engines, she found herself below a flight path leaving Heathrow, the busiest airport in the world.

Finally, she watched the navvies return — this time fewer in number, but equipped with great machines — to build the motorway that ran along her side of the valley, then leaped across it on concrete stilts.

A little while after its completion, men crossed interplanetary space and walked on the moon. Through television, this too could she watch.

Her life extended over a full century — one that marked greater change than any that preceded it. Surely the greatest change of all was to how, and how far, we travel. Perhaps most significant is how far we are now *obliged* to travel, and how much we have left behind.

Above her hearth was a sign that read :

> "For Pete's sake let's have no more progress."

1.3 The coming of the car

Fifty years ago, very few people could defend the claim that a car was anything other than a luxury. Given its tremendous cost, in a world still rebuilding after a cruel and destructive war, its ballistic rise to dominance seems one of the strangest things in a very strange era.

Dad's first car was an Austin Seven he inherited in 1955 when his father died. Mum insisted that we could not afford the expense it incurred. To Dad, it was everything. He simply didn't own anything else.

Men typically married soon after the war. The pipe and slippers routine was undoubtedly appealing after years of discomfort, away from home and family. The downside was that they enjoyed no adult interlude between service to the nation and taking on family responsibilities, especially financial ones. Not only did they have no money and no possessions, they had no space either. Until, that is, they owned a car.

Only now I have children of my own, do I understand why we boys were kept away from Dad for a while when he came home from work. He had an hour in the kitchen each day, in peace. Otherwise, the home was a roost ruled by Mum, and overrun by children.

The car was all the personal private space he had.

A similar situation existed at work. Around the office, he was subject to authority and accountable for every minute. Internet distractions would never have been tolerated in the 1950s. But once he left in the car for duties elsewhere, he was free. Trust replaced supervision. A measure of authority replaced subordination. His car was both mount and den.

It was also both a symbol of and a medium for his self-esteem.

It would be silly to pretend that the *only* value in a private car was emotional. It offered comfort while travelling, and shortened journeys.

Before he had a car, Dad would cycle the six miles to work and back each day. The train was available, but likely shunned for the expense. Though using a car undoubtedly cost even more, it was probably a little quicker and certainly more convenient. He could leave without delay, unconstrained by a timetable, and drive door-to-door, with no need to walk to or from the station, at either end. While his bike offered all these advantages, driving was far easier, and a lot less unpleasant in the wind and rain.

Soon, he would even be able to listen to the radio *en route*.

Longer journeys were different. For business trips and holidays, the train could offer still greater comfort, and its speed meant a far quicker journey overall. Driving for more than an hour or two palls anyway, and becomes tiring. On the railway, someone else drives. However, the price of four or five seats on the train made it cheaper to take the family by car, assuming you already owned one. Nationalization meant the railway was in no shape to invest, adapt, and compete properly.

My grandparents had always gone on holiday by rail.

We went by car.

Believe me, a family of five was a tight squeeze in an Austin Seven, and the journey was harrowing and far from safe. On the other hand, we were no longer tied to the railway resort. We were free to explore when we arrived, and could stay in unspoiled locations.

They would not remain unspoiled for long.

It would never make financial sense to run a car just for one trip a year. Yet, the more it was used, the more expensive it became.

There are those who like to argue that getting around by car is cheap. As we shall show in the next chapter, this simply isn't true, unless you choose to neglect everything but fuel. A car has to be financed, maintained, insured, and kept somewhere. Roads on which to drive must also be built, repaired, signed, and policed. These things are all every bit as necessary as fuel and collectively cost considerably more, as we shall see.

For those that could afford it, the *practical* benefit of owning a car then was merely a more comfortable and slightly quicker journey to work and back. Dad certainly had a hard time justifying the expense. On a modest income, meeting the cost of running a car certainly caused friction.

The willingness of so many men in his generation to make such an enormous financial commitment (often involving quite heavy debt) begs a bit more explanation than just an easier ride to work and a bit of privacy.

Clearly, the car is primarily an affectation.

We are not creatures whose behaviour is determined by mere pragmatism. There is comparatively little in what we do that can be explained without recourse to the emotional. The advertising industry could not function otherwise. Above all, we value our differences beyond anything, and each do our very best to preserve and express our individuality.

Clothing, for example, ranks high on our agenda.

Any new form of self-expression is likely to prove popular. The most recent addition is the mobile phone. There are literally thousands of models on the market, differentiated primarily by design — artistic, rather than technical. Failure to own something 'cool' makes you an outcast.

In the 1950s and 60s, inclusion meant owning a car.

The pressure to conform was well understood by manufacturers, who sought to promote the 'normality' of owning a car. They were perfectly willing to stoop to denigrating any who resisted as foolhardy oddballs.

Every mature male just *had* to have one.

Sadly, so did every immature male, though for different reasons.

As an ordinary lad, packed off to a school within a middle-class stronghold, I cared nothing for social status. I was used to "going without", and didn't mind being one of the few left behind by the class skiing trip.

What I did care about, hugely, was being able to get to the party. In a village, miles from the nearest town, you could perhaps *arrive* by bus, but the last one would leave embarrassingly early. Local railways were closed by then, though the same objection would have applied. There was already enough sprawl to make many destinations inaccessible by rail anyway, and enough traffic to make cycling dangerous.

Last, but by no means least, without a car it was virtually impossible to go out with a girl. Especially after dark.

No question, I wanted a car as soon as possible.

In all honesty, I can say that I have never associated my sexual prowess with a car. At no time in my life have I ever felt the need to drive something with an extraordinarily long red throbbing bonnet.

I can also say that I never saw the car as a place of opportunity for much more than a cuddle and a kiss. In the typical British vehicle of the time, there simply wasn't room for much more, without risk of injury, breaking something expensive, or getting stuck. (At least one couple in a Mini endured rescue by grinning firemen.) Maybe there's something in the *Kama Sutra* that would work, but I never found it.

Nevertheless, that private space the car afforded was highly prized.

Coward that I am, I choose to leave the question of women joining in as a matter for other people's Ph.D. theses. Suffice it to say that, for car (like cigarette) manufacturers, women are the new customers — *young* women.

It came as no surprise that the French were first to aim a car at women, portraying a liberated youngster driving rings around her father.

Although it is still quite rare to see a woman drive a man, things are changing. While my mother never drove at all, my niece is as likely to drive as her husband. Social changes seem to take about three generations.

Soon, there will be *two* cars outside every home.

Something else has changed, and at a deeper level, inside us. Two generations have now grown up having rarely, if ever, travelled by any means other than the car. My own children are still very young, but my nephew and niece are typical. Not until they were at university did they ever use a train. Throughout their early learning, travel meant one thing only.

To them, travelling by car remains *axiomatic*.

Anyone who has watched a child encounter the world for the first time, knows that reason plays little part. Experience is everything. Nothing is questioned, however astonishing. Objects and processes are fixed and timeless. They simply exist. Early learning about the world mirrors the learning of language. You need a vocabulary first, and begin with nouns. To understand the world, you begin with things. To understand things, you must learn what they do and how they are used.

Those brave souls who ventured so bold as to attempt the construction of artificial systems that might be labelled 'intelligent' were quickly beaten back. It is no easy task. Some responded wisely by asking precisely what it is they are trying to accomplish in the first place. Not all shared a common goal, but one sensible aim was the construction of machines that can respond to stimuli in the light of some consistent internal *model* of the world.

The term 'ontology' used to belong to philosophers, who like to debate the nature of existence. To the ancient Greeks, it meant "a theory of being". A modern philosopher would use it to refer to a systematic 'account' of existence. It has since been hijacked for a rather different purpose.

An 'ontology' has come to mean a set of processes and objects that a system may assume exist, at the outset. Processes and objects combine to form 'schemes' — the fabric of experience. A small child will get used to "being fed with a spoon" or "being put to bed". They will examine a spoon and explore a bed, and in doing so learn to recognize and use them, so that they may later conduct these processes themselves.

Apprenticeship is based on the same natural principle.

Childhood early learning is partly about establishing an ontology. Schemes learned early are not happily given up, because everything we know depends upon them. There is a counterpart in religion, which is often taught early for just this reason. A tenet is rigidly accepted, not merely because "it is written" but because bringing it into doubt threatens something *fundamental*, rather like messing about with the foundation of your house. It invites a catastrophic collapse of everything constructed above.

Everything new that we learn is necessarily an extrapolation of our world model. Learning is eased by the most extensive prior knowledge possible. We are ultimately bound by the constraints imposed on our early learning. When our experience is limited at any stage in our development, there is a 'bottleneck' effect, all the more profound the earlier it occurs.

Sad though it is to contradict even the best in science fiction, we will never comprehend the world of any real interstellar explorer, any more than a 19th Century peasant, who never left their hamlet, could ever comprehend 21st Century global commerce. We simply lack the relevant experience.

To inhabit a different world is what it is to be alien.

Challenging the axiomatic identification of travel with driving (or being driven in) a car can produce a variety of responses, from hypocritical sympathy, through confusion, to open hostility — exactly the sort of thing an alien might expect, should they query the behaviour of the locals.

A mathematical theory begins with axioms, which capture truth held from the outset, beyond doubt. From axioms, some mechanism of inference is employed to deduce further truth — typically, some form of logic. A 'theorem' is first postulated and then proved by applying inference upon both axioms and existing theorems, already proven.

And so the model grows.

If a single axiom is not *wholly* true, for *every* entity in the 'domain', then the entire theory may fail, regardless of the confidence placed in it.

We guard the fundamental elements in our model of the world more jealously, the fewer they are. Instinct tells us their value. Another tribe nearby, whose foundation appears in conflict with our own, will prove unsettling or even unacceptable, often with disastrous results. As our tolerance has grown, so has the meaning of 'nearby'. We are as wary now of the 'fundamentalism' in others as we are blind to our own. Perhaps some of *our* axioms lie in error. Perhaps *they* don't all apply to everyone.

But who would dare question them.

2

Ways and Means

2.1 Orientation

A matter of concern

Industry and society are changing at a rate not seen since the Industrial Revolution. In the rich world, factory work has given way to office work. Reasons for leaving the home to work are disappearing rapidly, as the means to be productive becomes portable and the internet remakes, and redefines, communication.

The daily dissolution of the family may soon be over.

Yet the need for human face-to-face contact has never been greater. It's just not with the same humans every day. People travel with their work as once did only their masters. Transport infrastructure has become the cardiovascular system of the modern economic body.

That system is currently failing, on both sides of the North Atlantic.

I live on what the Prime Minister Tony Blair described as the "techno-crescent" between Oxford and Cambridge. Yet there is no fast path between those two great, and lately industrious, university towns. The twin-track railway remains closed despite many attempts at reopening. Roads are congested. Journeys are tortuous; the cost in precious time too great.

The UK Royal Commission on Environmental Pollution predicted that congestion will continue to worsen, without major policy change, and that its effects will cause harm that would "not be acceptable environmentally or socially". They suggested an annual cost of around twenty *billion* pounds — a figure later reaffirmed by the Confederation of British Industry (CBI).

Yet this is only the *direct* cost, in lost production. The estimate does not include losses to the environment, business, leisure, or health.

Contrary to what you might think, we actually have *less* local and suburban mobility now than we did in the age of rail and bike.

Without fundamental change, our children may have less still.

For many years, surveys of the distribution of journeys by length neglected walking and cycling trips altogether. We make so many short journeys that they are hard to count. Our travel is predominantly *local*.

According to the UK Department for Transport, around forty per cent of all journeys are of under three kilometres. About four fifths are under ten, and nine tenths under twenty. We each typically make about a thousand trips a year (almost three a day), with an average distance of a little over ten kilometres. That distance has risen by almost a third over the last quarter century.[1] Curiously, the average journey *duration* has remained almost exactly the same, at about twenty minutes. The proportion of short journeys is in quite rapid decline. We make nearly a quarter fewer journeys of under a mile than we did fifteen years ago, while the proportion of journeys between fifteen and forty kilometres has increased by more than a third.

Distinctly noticeable by those of us long-in-tooth, is that the length of journey *needed* to accomplish many a task has increased dramatically.

While staying within a new residential estate recently, I found myself travelling ten kilometres just to post a letter. The same journey back home is of only a hundred metres. In my youth, it was thirty metres. Work and home are now typically separated by several tens of kilometres. Even schoolchildren suffer a doubling of their commute, as small and local schools give way to those with rolls tenfold greater.

Something is happening. Whatever it is, it would appear unsafe to consider transport in isolation from the pattern of development.

The question this chapter seeks to answer is why we seem no longer able to meet demand for transport, despite truly gigantic investment, predominantly in roads and cars. What is it about road transport that causes congestion, and why did rail and bike not suffer the same ailment?

---

1 The *total* distance travelled doubled over the final forty years of the 20th century.

Science typically proceeds by looking for properties that are shared among different phenomena. An *abstraction* then emerges that reveals, and helps explain, variation. Transport is no exception. In this section, we shall talk about transport in general, in preparation for the next, where we will examine each mode of transport in turn.

We begin by laying out the bounds of our concern.

Our interest here is limited to overland journeys of up to a thousand kilometres. Beyond that, and over oceans, flight may always possess an overwhelming advantage in speed. But then only rarely do we make such trips, even with our new-found wealth. Only in one way are they relevant here. Every long-haul trip requires a journey to and from an airport, which must therefore lie on some regional transport network.

It is useful to make two further distinctions; first, between the carriage of freight and of people. Here, we are concerned mainly with passengers, but must recognize the economic importance of moving freight. Most infrastructure construction has traditionally been justified by freight haulage. Passenger travel began as an added bonus, but has since risen to become both a prime aspiration and an economic necessity.

The second distinction is by journey length. We shall distinguish distances of up to six, from six to thirty, and thirty to a thousand, kilometres; referred to as 'local', 'district' or 'suburban', and 'regional', respectively. This allows us to consider appropriate options in each case from the outset.

One distinction we shall not be making is that of "public transport". Largely, this is just an unfortunate term that has risen in use so much that it has become hard to avoid. The most common interpretation seems to be in the form of a bus, which is usually privately operated, though often subject to public subsidy — for which, read public provision. Underlying all this is the cold assumption that the car is purely private provision. But the private car depends almost entirely on *public* roads, everywhere. Only a very small proportion of its 'track' is provided privately. Every car must also compete publicly for a space, in just the same way a passenger does on a bus.

Certainly, it is useful to distinguish between vehicles that are shared and those that are not, though we must take care. With regard to energy efficiency, for example, the degree of sharing is critical. While the bus may carry many at "rush hour", it typically enjoys low occupancy all the rest of the day. Any gain would also have to be set against a considerable waste when each passenger is forced to go where every other may wish, enduring delay at every stop. And, of course, the shared vehicle is *much* heavier.

A few passengers each travelling twice as far will save nothing.

The idea, promoted by many a 'green' campaigner, that we must somehow coerce, perhaps even compel, people to use "public transport" is rarely either clarified or properly justified. By far the 'greenest' mechanized mode is, after all, the humble bike, which is privately run and rarely shared. In contrast, an airliner — the *bête noir* of the 'greens' — is shared, just like a bus.

Sharing can often reduce transport cost and add value, but so can other things. While we shall speak again of public versus private *provision*, there are more useful ways to distinguish modes of transport.

Planning a journey

What factors influence our choice of how to travel?

Assuming we do indeed have a choice, some criteria are plain. First perhaps is *efficacy*. It must be possible to actually get to our destination. Next might well be *speed* and *price*. As in any market, people weigh value against price, according to their ability to pay. Given the choices people can be seen to make, it is abundantly clear that value is judged predominantly by door-to-door journey time. Time is money, after all.

From here onward, we shall use the term 'cost' to account for value and price in combination. Cost represents everything given up by the traveller in making the journey — predominantly, time and money.

We should be clear about the difference between the speed attained while on the journey and the *effective* speed over its entirety. Managing two hundred kilometres an hour may not be helpful if there is a long wait to depart, or if there are delays incurred in making connections. With aviation, for example, it is not unusual now for it to take longer to get to, from, and through, airports than to make the flight. Good "modal integration" can reduce the cost of making connections, given frequent service.

People loathe making connections, particularly when it involves changing mode. Connections increase journey time and decrease *predictability*, which is also highly prized, especially when the consequences of delay are serious; for example, when pay is docked, or a job interview missed. They can also increase price. Perhaps a vehicle must be parked. If multiple operators are involved, overall ticket price may rise.

The distaste for making connections may also be partly due to the disturbance. A change might make the difference between getting something done *en route* and not, making the journey time a waste. It always means getting 'settled' once more.

Published studies of the effects of connections are few and far between, though commercial operators are well aware of them. The best suggestion would seem to be that only *one in four* will remain customers on a given path following the introduction of a single change. Obviously, this may be expected to vary considerably according to the purpose of the journey and alternatives available, but it is clear there is a high cost perceived.

Something that is strangely often left out of our deliberation is whether or not time spent on a journey is actually lost or not. If one is free to engage in sleep, work, or pleasure, then value rises and the true cost declines.

The writer Arthur C. Clarke long ago asserted : [2]

> "… I steadfastly refuse to have anything to do with vehicles in which I cannot read while I am travelling."

Any need to pilot the vehicle yourself not only renders the journey time a complete loss but it taxes and tires you as well.

Rich people employ chauffeurs.

An increasing number of professional people wish, and often need, to work *en route*. Few will now accept being out of touch while travelling. Being in touch now means access to email and the internet, as well as a phone.

Comfort is another important point. People like to be pampered. They like being attributed *status* even more, perhaps with servants attending. Comfort should not be underrated as an influence over choice of travel mode, or even whether to travel at all. There was once a time when ocean liner and railway operators knew the value all too well. Pampering customers, raising their sense of esteem, once helped make fortunes.

Convenience is also highly prized. Easy access to a transport network raises the value of any property, private or commercial. Developers know

---

2 *Profiles of the Future*, 1962, p. 40.

this well and make hay out of it. Concern is rising, in Britain, over residential development lacking concomitant infrastructure. It is the market that is responding, despite the fact that the State has for decades been the sole provider of such things. With the city of Milton Keynes set to double in size, a group of developers recently *themselves* proposed a levy of ten thousand pounds per property built, to pay for transport infrastructure.

A path is hardly convenient if there is no way to tell that it's there. Following the arrival of the mobile phone and internet, we have high expectations regarding the provision of transport information. Ticketing also ought now to be simple and accessible. It need never interfere with travel. There is no reason anymore for delaying departure by selling tickets while boarding.

Customer service affects convenience profoundly. A traveller deserves to experience toilets somewhat cleaner than an animal enclosure, and to find help when service fails, or in an emergency. This might seem irrelevant to using a car on the road, but such is the illusion of 'private' transport. On a long journey, we still need toilets, directions, and help when we breakdown.

Closely allied to convenience comes *control*. Personally, I do not look forward to being pushed around in a wheelchair in my dotage. There is a flip side though. When someone else is steering, it is up to them to sort out the problems.

Which brings us to *safety*, and the *perception* of safety.

Obviously, safety ought to be right at the top in any rational assessment, but we are not creatures of reason, thank goodness. That would mean a world of the cheapest dull-gray paint colouring all we are and all we do. Perception is all, and we can choose where and how we look. Last year, I took the family on a plane. Channel surfing, as one does to avoid tedious adverts, I chanced upon an interview with someone who had lost his entire family on a plane. Poorly maintained wiring led to a tiny spark. The fuel tank blew the plane in two. His wife and children were at the back and experienced the kind of in-flight entertainment you prefer never to contemplate. He could not avoid knowing they would have survived several minutes of pure hell.

Safety, and the perception of safety, are very different. My perception of the safety of my family leading up to that flight was somewhat worse than reality. With some modes, notably those that use the road, it is the other way around. Safety perceived is much *better* than the real kind.

Where there is a choice, it would pay to have some common measure of safety. This is tougher than it sounds. Simply counting casualties pays no heed to numbers of journeys made or their length. It would be easy to harp on about the huge number of road casualties, for example, to which someone could easily retort "What do you expect? Most journeys are made that way!" Similarly, it would be easy to proclaim the safety of air travel by quoting the small number of casualties *per kilometre.* This would entirely ignore the fact that nearly all the risk is in take-off and landing, regardless of how far you go. (It would also ignore the very slim chance of surviving any accident.) With some modes, only a measure of risk *per trip* makes sense. We must ask just exactly what it is that matters.

If we take the time to look at the travel we do, two things are immediately apparent. First, the number of trips we make has stayed remarkably constant, varying in Britain by less than half of one per cent between 1985 and the millennium, for example. Looking farther back is unreliable because we were much poorer and relied far more on short trips, walking and cycling. These were notoriously hard to count. As mentioned earlier, many surveys after the war ignored them altogether. In any case, it makes sense that the number of trips remains the same, simply because the tasks we travel in order to perform remain much the same. We go to work and take the children to school. We may spend more, but go shopping no more frequently than our parents did. We may go on holiday more often, but still comparatively rarely.

Second, we are using our greater wealth and leisure time, not to travel more often, or to spend more time travelling, but to travel *further* each trip. While the number of trips varied by at most eight per cent in the last quarter of the last century, and their duration stayed roughly the same, their length rose forty-four per cent.

Because we are ignoring aviation here, and because the number of trips we make has varied so little, we shall opt to compare the safety of modes via their casualties per kilometre, or rather per billion passenger-kilometres (bn pass-kms). We shall count only the dead as casualties. Only deaths that result from injuries sustained within thirty days of the accident are included. (There is some consistency over this in published statistics.)

We need to consider each mode in isolation. This is difficult where the major cause of accidents is some other mode, in which case it makes more sense to consider casualties *inflicted* when discussing the mode responsible. Where relevant, we shall distinguish casualties incurred on one mode by another.

We should always be very wary of statistics, especially when published by anyone with an interest in what they are supposed to convey. Elected regulatory authorities, as well as commercial operators, have an interest in seeing casualties appear low and falling.

There are also other factors, sometimes hidden. It is not always the users of a transport mode that pay the price of increasing its safety. Sometimes others have to curtail their activities, or suffer some other imposition. If a boasted fall in casualties is achieved only by removing people from the threat, rather than the threat from the people, it hardly has the same value.

While we tend to put it to the back of our mind, as individuals, safety is a paramount concern within the broader view we set out to take.

Another issue is our *health*, which also turns out to be strongly related.

Suppose it is true that, in the words of Frank Borman, after flying around the Moon, "man can now do anything he wants to, technically". One need only explore science fiction to see the kind of transport system we might construct if technology really was no limitation. Perhaps the most obvious example is teleportation — the ability to transmit *people* in the same way we now send messages and signals, by wire or radio. My favourite invention is the "moving way" of Arthur C. Clarke, in *The City and the Stars*, which depends on a material that behaves like a solid in one (vertical) plane and a liquid in another (horizontal). Imagine being able to walk out across a river. The water moves quickly near the centre but slowly near the edge, making it easy to board and navigate. Such 'rivers' could easily support junctions and allow a complete network to be established. No vehicle would be needed. They could fulfil every aspiration we have thus far mentioned.

One penalty of either system is that we would get no exercise at all. Our health would suffer. Travel has always been an important source of exercise. When work moves from the physical to the mental, as it has for many of us, getting about easily becomes the *only* exercise. Losing it would require us to face a stark choice find significant extra time in the day for deliberate exercise, or face complete physical atrophy.

There is clear evidence that our children have already lost that source of exercise. In Britain, the proportion travelling independently to school fell from four out of five down to just one in ten between 1970 and 1990, and then to just *one in twenty* by 2006.[3]

Once they walked or cycled, now they are driven.

We shall return to the subject of child health. Suffice it here to say that it is in a state of collapse, associated with a frightening increase in the proportion who are overweight and even clinically obese. There is clear evidence that the predominant cause is a profound reduction in exercise, and not eating to excess. Children in the 1950s actually ate more.[4]

Perhaps we don't always pay much attention to our health when we choose how to travel. But that isn't to say our children won't when they grow up. Such is the catastrophe that health is currently suffering, only those of our children who *do* take it into account may survive us.

We must hope they can afford to pay for the care of those who don't.

Our choice of transport mode also affects the health of others.

Walking and cycling give rise to no significant pollution. All motorized modes currently cause considerable air pollution, known to have serious consequences to respiratory health. Other effects range from headaches and loss of concentration to cancer.

Between twelve and twenty-four thousand deaths each year, in Britain, are directly caused by air pollution, according to one recent estimate.

In summary, when we plan a journey, we principally seek out the way that minimizes the door-to-door journey time. Any need to make detour or connection is far from popular. We may also be influenced by predictability, comfort, convenience, and the degree of control we retain. Rising in importance is the ability to stay in touch and use our time *en route*. Assistance along the way matters, and poor service doubtless deters. Those of us who

3 The same charity reporting this (*Play England*) added that two out of three children, aged eight to ten, now never visit a park to play, unaccompanied by an adult.

4 UK Medical Research Council National Survey of Health and Development, 1999.

are a little more circumspect may care about our safety and health. Some may even have regard for the safety of, and impact upon, others.

Concern over transport *capacity* we justifiably leave to 'planners'.

Co-operation and social value

Each of us would like to believe ourselves an altruist, making our decisions with regard to the greater good. The truth is that we tend more often to behave as opportunists, quietly taking the path best serving our own primary self-interest — at least, when no one else is looking. To be fair, our choice is often made without ill intent, though it may still have that effect.

Joining an already congested conduit causes its speed to fall a that bit further. Everyone passing along any such path is thus helping to slow the passage of others. They impose a *cost* upon each other. Conversely, joining a lightly-used service can aid its viability, reducing the cost to everyone else. Mutual *benefit* is then conferred.

Our choice of transport mode can also have several effects on fellow travellers, bystanders, and even on people far removed. Overall, each travel option can be said, on balance, either to promote society or to deny it. In choosing, we might weigh the *social value* of each — if, that is, we wish to pretend to a little altruism.

We might consider how tolerant each option is of other modes. One mode can interfere with another. Interference may be direct, perhaps by making it dangerous to share a common path, when no other is available, or indirect. Perhaps new controls and signalling are necessary for one, making the other slower.

It really isn't very social to drive others off the road, but the implications may run deeper. It may be that the mode affected carried more people along the same route, and at a lower cost. Perhaps it created less pollution.

Equity of access ensures inclusion, which is perhaps the hallmark of a civilized society. To deny mobility to the young, for example, is arguably to damage their development. Discrimination now often faces law and litigation. Both the elderly and the disabled have the vote and will not accept exclusion. Yet we remain free to choose a mode of transport open to only some of us. Inclusion, it appears, is not *our* responsibility.

We might hope for co-operation between transport providers. After all, two connecting services usually have a higher value than either would independently. Such synergy is easily disrupted by either market or regulatory structure. For example, two services might be provided by different operators, in competition. While they would both gain by co-operating, they fail to do so, perhaps out of mistrust.

They suffer what is called "prisoner's dilemma", because it is similar to the situation where two prisoners are each offered the chance to give evidence against the other. Their best joint outcome is to co-operate with each other and remain silent. The best individual outcome is gained by talking while the other remains silent. Strategy, under such circumstances, has been studied by contriving games. So-called "game theory" helped to determine policy during the Cold War.

In Britain, rhetoric has been plentiful with regard to transport integration, but we still suffer what is arguably the most *dis*integrated system in Europe. In 1947, when the entire railway system was 'nationalized', the British Transport Commission found itself the owner of an enormous number of horses and carts. Most railway companies guaranteed to deliver you, or your package, door-to-door and had the means to do so. Such commitment, organization, and service, soon disappeared.

Now that new private train operators are seeking to 'grow' their business, there are signs they are waking up to the primary aspect of customer service, namely helping their customers *complete* their journey, at either end. Some, at least, have plans of their own for modal integration.

What planning authorities in Britain have failed to do in sixty years might at last be brought about by business. Time will tell.

Shared modes, like bus and train, are vulnerable to other shortcomings, and any idea that business is automatically free of disorganization is sheer delusion. Timetables are sometimes a well-kept secret. Getting up-to-date information can be difficult if not downright impossible. Obtaining a ticket is still often allowed to delay, or risk missing, a service. Operators have been slow to properly exploit the potential of information technology.

One great strength of shared transport is its ability to deliver lower cost with high usage and high vehicle occupancy. Conversely, its great weakness is its vulnerability to poor planning and antisocial patterns of development

— in a word, *sprawl*. Once population density falls below a certain level, the additional delay and difficulty in getting to and from stops and stations makes the whole idea unattractive, compared with private transport.

The effect is a sharp decline in viability beyond around two thousand people per square kilometre, or about ten homes per hectare.

One reason why you will encounter vastly more shared transport in Europe than in America is that the combination of 'unlimited' space, relaxed planning regulations, and unprecedented growth, has led to very low density development in the US since the War. Of nearly five hundred urban areas in the US, only eighteen come close to affording "mass transit". (New York is the notable exception, where the *majority* commute by shared transport.) In Europe, a high proportion of the population live in towns established centuries ago, and which retain comparatively dense housing.

To provide viable and competitive shared transport within sprawl takes careful planning, effective modal integration, and heavy public subsidy.

Sprawl is, in fact, predicated upon universal access to a private car, although, on a limited scale, a bike might remain an option.

Here, we shall henceforth term the alternative *focal* development.

We are social animals. You have to ask what gives us the opportunity to meet and get to know our neighbours, given that we no longer toil together in the field. When we separate work, amenity, and residence, then the latter becomes little more than a dormitory. Our 'home' becomes disconnected from our society, and is thus diminished. To be surrounded by strangers is unhealthy. Alienation breeds contempt, and contempt breeds hostility.

People have a natural tendency to bond with a place. Evolution led us to *own* the space in which we live. A village belonged to villagers every bit as much as they belonged to it. We had been living that way for at least five hundred generations, since agriculture began after the last ice age.

Sprawl has ended this. It is impossible to identify with a place when it lacks both focus and amenity. There can be no sense of community where nothing of importance is shared. There is no 'glue' and thus no cohesion.

Annihilate the social, and one must then expect the antisocial.

Young people, in particular, give vent to ancient feelings they could not hope to express and seize the territory in which they find themselves. Urban graffiti is rather an elegant response, and arguably justifiable. Far from visiting damage on what is supposed to be communal, they are attempting to make it their own. Sometimes, the feeling leads to a profound insecurity, quite appropriate to any animal finding itself in a foreign domain. A display of strength and hostility is called for. When vandalism fails to satisfy, there lies only the path of self-destruction, with no care for who else comes too.

It is very hard to see why sprawl is still being allowed to replace viable communities. As is so often the case, we must choose between conspiracy and incompetence for an explanation, with the latter requiring fewer and easier assumptions. In an article on the economic opportunity in pursuing energy efficiency, Amory B. Lovins, a very successful consultant, wrote : [5]

> "If federal, state and local governments did not mandate and subsidize suburban sprawl, more of us could live in neighbourhoods where almost everything we want is within a five-minute walk."

You cannot separate land use and transport. One infers the other. *Focal* development can both reduce transport demand and lower its cost.

Perhaps it is neither conspiracy nor incompetence that gives rise to sprawl, but the lack of any structure, market or regulatory, that is capable of bringing forth anything better. One leading authority on urban architecture declared sprawl the result, not of bad planning, but of *no* planning.

Across Britain, local amenities, like shops and post offices, are closing. The result is a lifestyle typically robbed of all opportunity even to discover your neighbour's name, still less share anything in common with them.

Zoning of development compels the separation of residence and workplace, and thus an extended journey between. In the final decade of the last century, the number of commuting journeys of more than fifty kilometres grew by nearly a third. Residential and workplace 'neighbourhoods' are now entirely different. No wonder burglary has increased: who can tell burglar from neighbour? No wonder a "long hours" working culture has developed: if one spends more time at work than at 'home', there is a better chance of friendship, and perchance a 'relationship', of a limited kind.

True we are all different. A lifestyle that would be hell to us might be heaven to someone else. But then there is all that new hostility around to account for. There is also, not so much a rising tide, as a tidal wave of

5 *Scientific American* (September 2005).

divorce, emotional and mental illness, self-harm, and suicide. Nuclear weapons cannot defend us against this threat, though it is perhaps a bigger one than fascism, communism, or religious 'fundamentalism'. It may be more dangerous even than global warming and climate change, and has crept up on us with equal speed and stealth. The reason we have not seen it coming is that so many have never known that which has been lost.

They can only experience the consequences of losing it.

Society is not something that simply happens miraculously. It grows from shared experience, watered by culture and tradition. Growth requires stability and investment, as does that of an economy, which is, after all, a mere facet of communal life. Society cannot arise out of nothing.

We are what we share, whether we like it or not.

Our *habitat* is as important to us as the Indonesian jungle is to the orang-utan. Unlike theirs, ours is ultimately the product of how we choose to live. Ultimately, the choice does belong to us, despite how it might seem.

We reap only what we sow.

Automation

A second industrial revolution is underway, more thorough than the last. This time, *every* phase of production holds the possibility of mechanization, even the recovery of raw materials. Automation can now refine materials and fashion components. 'Lights-out' factories can assemble products. 'Just-in-time' (JIT) production can even configure each individual product according to a specific order, without any human intervention. Retail is moving from the high street to the automated warehouse and internet site, where a much wider choice can be found, with no need for the shopper to travel.

Goods can move directly from factory to consumer.

Employment has already moved from field to factory to office within four generations. Now the continuation of work, at least as we have known it, can be brought into question. As labour declines, capital must take over as the predominant means of recovering an income. It may be that a huge underclass may emerge, dependent on redistribution, but, one way or another, the age of drudgery is coming to an end.

It may be said that all civilizations have depended on some form of slavery. In the autonomous computer and robot, we at last have a form that is morally acceptable, ... and dramatically more productive.

There are two obstacles on the path to Paradise.

First, there is the problem of population and resources. Whatever economists might say, production remains dependent upon energy and raw materials. Nowhere near enough of these exist to offer the entire global population anything like the rate of consumption prevalent in Europe or America, never mind anything beyond.

There can be no place for greed in such a Paradise.

This obstacle can only be removed over coming generations via the conquest of either greed or overpopulation, or perhaps both.

Second, we cannot automate what has become the primary transport network.[6] Road transport can never be rendered autonomous. However, building an automated transport network is well within our capability.

There is no place in our Paradise for the car or truck either.

2.2 Options

It's our choice in the end

Having spelled out all the many things we might take into account when choosing how to travel, it is time to consider our options.

At the personal level, we may not have the luxury of choice at all. As we have seen, most of the development in both Europe and America since the war has been predicated on universal access to a private car. There is rarely any alternative. Rural areas in Western Europe, especially Britain, Germany, and France, were once very well served by the railway. The decline in rural populations and the rise of competition with bus, lorry, and car, have largely removed that option. Buses have since become scarce. Choice has again declined. Many must now run a car, beg a lift, or stay at home.

6 Do not believe anyone who promises a computer that can drive a car or truck. Such promises have been made for forty years but remain utterly unfulfilled.

Assuming we do actually live in a democracy, and that we care about liberty (*i.e.* choice), there remains the possibility of planning a better habitat for ourselves. We might prefer to reduce our need to travel by rendering amenities closer to our homes. We might also wish to reduce pollution, especially where we live, work, and play, use less energy, derive it from a sustainable supply, and reduce the cost of transport. (Recall that the cost of transport is measured in journey time, as well as ticket price.)

Above all, perhaps, we might wish to reintroduce choice.

Since planners are ultimately answerable to an electorate (though the path may be long and tortuous), we can at least hope that the need for choice, and all factors that affect personal preference, will influence planning.

Of course, this assumes that planning does indeed take place.

Throughout this section we shall employ the same set of rules with every mode, to enable a fair comparison, especially of capacity. Although, in each case, some deviation may be justified, we shall generally assume :

— a minimum aperture (lane width) 50% wider than the vehicle/person

— flow to be 'duplex' (in both directions, requiring two lanes)

— an occupancy of one passenger per private vehicle, or 50% if shared.

A rough measure of capacity in terms of people per hour (pph) per metre-width of aperture (pph/m) can be calculated, given the typical speed and the width and length of the vehicle (or person). Aperture (or lane width) matters for three main reasons. First, it is vital to make the most of narrow streets and alleys. Many urban environments evolved for passage by horse and cart at most, and offer little room for anything more. Second, the wider the aperture, the greater the space taken up. Land is a major cost when establishing any transport conduit, with regard to purchase, construction, and maintenance. Third, wider conduits are more difficult, and expensive, to cross. Other problems such as drainage can also emerge.

Walking and cycling

Walking might seem slow, but *not* if one considers it properly. Living in a village, I can reach shop, post office, railway station, and church, within a five-minute walk from my door. I can travel without delay, and without making connections or stopping to collect or drop off others. To walk is free, if you disregard shoe-leather and a hearty breakfast. It is also, intrinsically, almost totally safe, open to nearly everyone, and restricts the liberty of no other soul, in any way. Only very modest investment is needed.

Any traditional village since agrarian settlement began, around eight thousand years ago, provided the ability to make almost all trips on foot. It was a good job too, since a horse was a luxury, affordable only to the gentry. Ordinary folk fetched water and food, and visited church and family, *à pied*. All necessary amenities, including the family support so vital for child care and in times of crisis, were attainable this way. Because of the need for proximity, the same set of people would share every amenity — the same shops, workplaces, school, church, reception hall, and pub.

Most of this has gone from the typical rural village, but survives in many a small town. The point is that walking can not only work in a habitat that is dense and strongly focal, but can also promote society and community. Its cost, like its efficacy, depends strongly on the nature of the built environment. We seem to be happiest with a journey time of up to twenty minutes. If the majority of amenities can be reached on foot within that time then most of us are quite happy to walk.

To be able to travel further afield is obviously desirable, but remember that the vast majority of the trips we make are local. The ability to make these on foot is vital in keeping the cost of transport down. In the poor world, where little else is available, people spend just three per cent of their time travelling. In the rich one, we spend more than a quarter of our life, and nearly a quarter of our income, on the road. Perhaps this is largely just how we choose to spend our extra wealth.

But then we must ask how often we have any option but to drive.

Nature intended us to walk, and we are beginning to learn the cost of living in conflict with the way we have evolved. Walking is not just the cheapest option, it is also extremely healthy. An hour's brisk walk typically burns off four hundred calories. Just three ten-minute perambulations a day can halve the risk of heart disease and stroke. Walking can lower blood pressure and cholesterol, and help boost the immune system and maintain body chemistry. It can also relieve anxiety and stress, especially when in the midst of natural surroundings. For me, this is particularly true of woodland and water.

Walking is in severe decline. According to the British Department for Transport, the number of trips made on foot fell by nearly a quarter over the final fifteen years of the last millennium. Only one in three trips are now made on foot, accounting for just one in thirty of the kilometres we travel. Shanks's Pony is joining sailing ships on the dusty shelves of history. We are advised to take around ten thousand steps a day, but now manage an average of only three thousand. According to the National Health Service, only a fifth of us now get sufficient exercise. Half our children engage in less than two hours of physical activity per week. In my childhood, I did that much each *day*. The World Health Organization recently concluded that indolence is now a bigger threat to health in Europe than smoking.

Walking is actually *necessary*. Deny that at your peril.

Three teachers stand out in my memory. All for something provocative they said. 'Snowy' White was my geography teacher — a bright-eyed walking inspiration in his eighties. For a toffee, he asked us what was the most important mechanical mode of transport. Being young and male, you can guess our answers. One by one he carefully demolished them, with argument of an elegance now so rare.

I remember the delight in such a surprising answer.

A conduit just 1.5m wide should allow around 3,000 pedestrians to pass in each direction per hour. This assumes a half-metre stride, and a half-metre width, paced at 3kph. A cycling conduit need be no wider (unless it is intended to accommodate large trailers) and could carry ~2,500 people per hour (pph) at 15–18kph. This leaves figures for capacity per unit aperture of about 2,000 and 1,700pph/m, respectively, but with a range and speed *fivefold* greater for the bike, at almost no extra charge.

The figures say it all. Snowy was right.

Cycling offers an effective complement to foot-slogging. It makes that trip to a shop, workplace, or school, on the other side of town, simple. I ride the six miles into town at least once a week, partly for pleasure, partly for exercise, but mostly for convenience. Only the worst wind-driven rain puts me off. As with walking, I can go where I like, when I like, without delay, connection, or account of other passengers.

It is also possible to carry quite a lot. My personal record is a lawnmower two miles, though I admit I would prefer other means next time. Children offer no problem. We carried one on a crossbar seat, and one on a trailer-bike. Where safe routes to schools exist, as many as two of every five pupils arrive by bike, offering tangible relief from morning congestion.

Many an anecdote is told to supposedly support an objection, in which there is little truth. Hills, they say, are the reason why few ride a bike, or why it is irrelevant in some given area. True, that cycling may be more popular in the Netherlands because it's so flat there. But then Dutch bikes seem usually to lack a wonderful invention — *gears*. Just like a car, you change down when riding uphill. A recent fashion for "mountain bikes" has seen gears so low that you can ride uphill with less effort than walking. In any case, a bike can always be pushed. Then there is the reward. You get to ride down the other side with no effort at all. For a thrill, the sensation of speed on a bike down a steep hill beats everything bar a bob-sleigh.

We ought not to forget the importance of *fun* as we grow up, or old.

Another myth is that cycling is open only to the young and athletic. An elderly neighbour of mine continued to ride when he could no longer walk. Many are the disabilities that exclude you from driving a car. Few will deny you a bike. In any case, there are now solutions for those unable to pedal. A revolution in battery technology has increased its power-to-weight ratio dramatically. Electric bikes ('e-bikes') can only get more attractive from now on, making cycling open to almost everyone.

Finally, there are weekend letter writers who claim that cyclists regularly strike fear into, and injury upon, pedestrians. I make frequent use of shared paths and have never witnessed conflict. Obviously, the antisocial appear on a bike on occasion, but the damage they can do hardly compares with that caused by a careless driver.

Ever since mediæval times, when witches were put to the torch, humanity has defied its title by relishing a pogrom or two. A tainted heart with a gift for rhetoric has rarely failed to find an eager mob. For an example of thoroughly hate-filled invective directed against the humble pedal-fiend, consider the following, from an article in The (London) Times entitled *What's smug and deserves to be decapitated?* by diarist Matthew Parris : [7]

> "A festive custom we could do worse than foster would be stringing piano wire across country lanes to decapitate cyclists."

One such attack was reported the very next day. Luckily, the hapless, and no doubt innocent, victim survived, though not without injury.

Another contention of those oddly hostile to cycling is that they do not pay for the road they use. Studies have repeatedly shown that 1) cyclists would willingly pay for their own causeway, properly separated, and free of the usual sunken drains and potholes, and 2) the necessary fee would be a tiny fraction of that needed to provide and maintain paths for motorized traffic. The problem is that the cost cyclists would incur, even so well-served, would typically be significantly less than the cost of collecting the fee.

FIGURE 2.1 A safely separated cycle-way.

Drivers of motor traffic do, in any case, have something to gain by sharing their causeway: a bike consumes far less space than another car.

Cycle-ways are extremely effective in using the minimum lane width to deliver the maximum mobility, all without causing any pollution, of any kind. Once built, they are typically well used, provided they are both continuous and safe. Many cities around Europe have adopted the bike to the profound benefit of both environment and mobility. The example has been set, and is there to behold.

Safety requires the separation of a cycle-way from both motorized traffic and pedestrians. Separate crossings will avoid interference and yield up

7 December 27th 2007.

the true potential of both foot and bike. Figure 2.1 shows a typical Dutch cycle-way, complete with its own signalling.

Sadly, cycle-ways remain a rarity in Britain. When they are found, they consist of mere paint on the road. They receive the lowest priority for a share of available road space. Hence, when the road narrows they suddenly disappear, yielding to motorized traffic (Figure 2.2[8]).

FIGURE 2.2 A cycle-way in Britain.

Cycling is roughly *five times* faster than walking, yet requires an aperture no greater. A bike is just longer, not wider. While its speed brings no greater capacity, it does translate into a fivefold greater range. A half-hour cycle ride will take you nearly ten kilometres — a half-hour walk less than two, with the same effort — all without consuming anything more than some extra pudding.

When you look at how the proportion of trips we make varies with distance, you discover a curious thing. Those who prefer a bike use it in very much the same way others use a car. The 'profile' is much the same. For the majority of short trips, just too far to walk, cycling offers a genuine alternative.

While the value of the humble bicycle is clear for trips between villages and across town, there are many who believe it has the potential for inter-urban journeys also. In Britain, SusTrans (an abbreviation of "sustainable transport") has led the construction of the 16,000km *National Cycle Network* in Britain. Three quarters of the population now live within three kilometres of it, and usage is steadily growing. They have also championed *Safe Routes to Schools*, enabling and encouraging thousands of children to regain control of their daily travel.

SusTrans also advocates one other rôle for the humble bike, in their *Safe Routes to Stations* initiative. The bike presents great opportunities for

8 In colour, you would see a green cycle-way end abruptly where the (red) bus lane begins.

modal integration because of its small size and convenience. Parking is cheap to provide, and, unlike using a bus, there is no need to worry about connecting services. Indeed, there is little reason not to take your bike along with you on the train, especially if it is a compact folding model. It can then provide the means to complete the journey at *both* ends, and at almost no additional cost, regardless where you go.

Commuting would still seem to offer the greatest potential for cycling. In London, the Congestion Charge[9], and bomb attacks on the Underground (railway network), have encouraged many to take to a bike, though support with cycle-ways and bike racks is almost non-existent. Between 2000 and 2006, there was a 70% rise in journeys pedalled. Sadly, this saw a nearly threefold increase in fatalities, while the national figure rose just 10%.

Internationally, statistics suggest that the more cycling goes on, the safer it becomes. That does not yet seem to apply in London. Concern over safety is arguably the principal deterrent to any further increase.

In 2004, UK Department for Transport figures infer thirty-nine dead per billion kilometres cycled — nearly *ten times* the figure for the car. Fifty-two people die just walking, by the same measure. The two transport modes that are intrinsically the safest become far and away the most dangerous in the presence of the automobile, which accounts for almost all injury.

Small wonder people seek refuge in a car.

Cycling has similar health benefits to walking, and more besides. Recent studies have revealed the importance of periodic *vigorous* exercise. Cycling was part of most people's lives for generations after its inception. Sadly, it is suffering a decline similar to that in walking. In Britain, the average distance pedalled fell by a quarter over the final quarter of the last century. The situation is rather worse than it might seem, however, as the number of *trips* undertaken by bike fell by *half.* It may be that cycling for sport is rising slowly with wealth, while cycling for practical purposes is declining rapidly. Oxford was once a city famous for its bikes. If the current trend is maintained, they will be all but gone within a decade.

It is not the same the whole world over. Where opportunity and encouragement are found, people are making full use of both foot and

9 A tax levied when you bring a car into the centre of the city.

bike. Copenhagen is a fine example, and able to boast the largest pedestrianized urban area in Europe and excellent provision for cycling. More than a third of its people arrive at work on a bike, and nearly another third on foot. Four fifths of trips made around the city are made on foot. Three times as many trips are made by bike than are made by bus.

Copenhagen is not alone in civilizing its transport. Nor does the car driver lose out, strange as it may seem. Recently, I spent a week in Eindhoven — an industrial town in the Netherlands. Everywhere, there are cycle lanes, fully separated from all other forms of traffic. Pavements are wide. People readily avail themselves of the opportunity to walk and cycle in safety. Because space is properly allocated for foot and bike, rather than exclusively for bus and car, traffic of *all* kinds runs freely.

FIGURE 2.3 Bikes too afford a little self-expression.

Where the will is there, it *can* be done.

Beyond any concern over throughput and avoiding congestion, cycling is fun and often exhilarating. It may well be found as good for mental health as it is for physical health, assuming the two can ever be separated. Even the less fit can join in, with assistance from an electric motor.

Perhaps, in the end, what distinguishes the bike is not its human power but simply its minimalism. Any vehicle that can fit on a cycle-way, and be limited to a safe speed (say 30kph), might be admitted as its complement. Its competition — the car — is merely something prone to an excess in both size and mass. This began with the need to accommodate an engine that could pull its own weight, and that of its fuel. New, much lighter, electric motors, batteries, and fuel cells, may well bring a spectrum of new minimalist vehicles that are smaller and weigh less than their passengers.

But the ordinary humble pedal-cycle will be hard to beat.

Bus and bus-way

What distinguishes, indeed defines, the bus is its ability to make frequent stops to collect and deliver passengers. In London, bus-stops outnumber railway stations about a hundredfold, with the result that one is likely to be much nearer the start and end of each journey — perhaps close enough to walk. On the other hand, this very feature is its greatest weakness. The bus is horrendously *slow*, with an effective speed way below that of a bike. Where it is used for direct services, with few stops, it loses its big advantage but remains far slower than rail, which then becomes viable.

Where sprawl has been allowed to develop, rail service is typically impossible because of the large number of lightly-used stops required. The bus is the only form of shared transport that can begin to match demand directly, albeit poorly. An indirect solution may still be possible, however, by combining bike and rail. Where an entire city has been predicated on the car, even the bus will fail to serve. As the number of stops rises towards the quantity needed, journey times fall below what would be acceptable.

Remember that, even if you *can* walk to a bus-stop, the service may not take you where you want to go. Within sprawl, destinations are many, and hubs are few. Connections will render the journey slow and tortuous.

Bus, bike, and car, are typically used with the same distribution of suburban trip lengths. In London, where there is real choice, rail is often the preferred option for journeys of more than a few kilometres. Below that, cycling, driving, or hopping on a bus, seems each to fulfil a similar function, though with varying appeal.

The popularity of the bus, though in decline, is surprising, given that it is far the slowest of the three, and arguably the most expensive, as we shall see. It may be solely down to physical ease, compared with the bike, and the convenient proximity of a bus-stop. For some, it's the sole option.

You certainly won't choose the bus if you're in a hurry.

In cycling to, from, and through, Oxford, I am regularly compelled to leap-frog buses. A bus stops to collect passengers, and fares for at least half of them. I overtake. Then it overtakes me. This is repeated two or three times, before I *stay* ahead. This, for most urban cyclists, is simply a matter of regular experience. Cycling is evidently significantly faster.

As we have said, it is in the nature of a bus — its very strength — to stop frequently. The effective speed attained thus requires a little thought. One thing is obvious from the start: the 'dwell-time' at each stop is critical; the more people that use a service, the longer the dwell-time will be.

Let us plan a bus route, with appropriately frequent stops. It seems reasonable to assume passengers are happy with, say, up to a five-minute walk to a stop. Therefore we place stops at intervals of around a ten-minute walk, say every three hundred metres. A typical bus is able to travel between these stops in around thirty seconds, allowing for acceleration and deceleration. This assumes some concern for speed limits, other road users, and passenger comfort. Boarding, however, takes a time proportional to the number of passengers waiting. The average bus journey would be of around six such stops, and we would like our bus to run at least half full. Hence, we would hope for five or six passengers boarding at each stop. After a while, a "steady state" will be reached with an equal number getting off and on.

A major distinction must be made, between whether or not ticketing is allowed to delay departure. (There is also a question as to whether passengers are allowed to find and occupy a seat first. It is rarely so, despite the compromise with safety.) Let us label the case where boarding alone delays departure *Case A* and the alternative *Case B*. Note that the truth in any particular area may be somewhere between the two. While it is often *possible* to buy tickets 'off-line', not every passenger may do so.

Allowing ten seconds for five passengers to board, the effective speed with *Case A* works out at 27kph, which is respectable and would compare favourably with other ways to get around town, notably a bike. If we allow ten seconds *per passenger* in *Case B*, however, the figure halves to just ~13kph. Here, we have an explanation for my experience of leaving behind buses when cycling into town each week.

These figures would not be so bad but for two things. No account has been taken of congestion. Secondly, neither buses nor passengers are evenly separated. The arrival of both will be subject to *bunching*. Buses bunch because of the effects of variable delays caused by route congestion and passenger boarding. We must now wait for the bus since its arrival has become unpredictable.

Waiting at a bus stop is particularly miserable. Comfort and weather protection is usually minimal, at best. Information is usually absent altogether. You rarely have any idea how long you must wait or even if the service is still running.

I once spent a week evaluating the time I would have to allocate to use a bus rather than the bike on my way to work. I recorded the delay each morning. To be sure of arriving on time, I found I would have to allow thirty-five minutes for a journey timetabled to take seventeen. (The bike ride took just fifteen, including a long steep hill.)

The effective speed of bus travel is *dramatically* reduced by bunching. With on-board ticketing, we should expect it to fall to a mere 6kph — just twice walking pace, or around a third the speed of a bike. The truth, as observed, is only a little better. In both London and Oxford, the average road-speed of buses has been measured at only 13kph. This does not account for time spent waiting at a bus-stop. The London Evening Standard used Transport for London information to estimate bus travel time around the city centre along twenty-five different routes.[10] Only eight were faster than walking, despite many dedicated bus lanes.

Nor is the bus cheap. My personal worst is paying £1.20 to travel three kilometres at just 6kph — and that ignores the wait.

Let us be charitable and consider a bus service with *off*-board ticketing and automated on-board ticket checking. The latter might most easily be achieved with a "smart card" that is consulted when both entering and leaving the bus, ideally by radio. A debit can thus be made according to the journey made, taking into account when it is made. Some schemes even take account of other journeys in order to offer an appropriate discount for frequent use or promotional purposes.

By avoiding ticketing delay at stops, the bus might deliver an effective speed (accounting for waiting at the bus-stop) of around 12kph, with a consequent 26,000 people moved per hour (pph) along a dedicated conduit. Even if that conduit is completely free of all other traffic along the entire route, it will only avoid delay caused by congestion, and *not* bunching.

Buses need a wide aperture through which to pass, which further dilutes what would otherwise be an impressive flow, particularly for ancient towns

---

10 *Return of the Traffic Jams*, 27th June 2006 (three years after the introduction of the Congestion Charge).

where space is at a premium. A figure of about 3,500pph per metre aperture (pph/m) results for *Case A* (no fare collection at stops). (Recall that this is duplex flow, *i.e.* 3,500pph in each direction.)

For buses that both bunch and collect fares at stops (*Case B*), capacity falls to roughly that for the bike — just ~2,000pph/m. In either case, when they operate along with other traffic, along *any part* of the route, the applicable figure is likely to diminish significantly because of congestion. The effective speed of my bus to work, allowing for both delay and unpredictability, due to both congestion and bunching, only just exceeded walking pace, which infers a capacity of just 1,250pph/m — a third less than a cycle-way.

This final figure is the realistic one for peak-time, when most people want to travel. Most buses still collect fares *en route* and share at least some of their path with other traffic. They will thus suffer its congestion.

They are also capable of *causing* that congestion. Being rather large, and stopping frequently, they could hardly be better contrived to do so.

Thus far, we have assumed our buses travel through homogeneous traffic. Unfortunately, this is rather fanciful. The majority of routes are radial, feeding into a town-centre, where buses become concentrated.

Oxford is a fairly typical town, which adopted a strategy based entirely on the diesel bus. It began with the introduction of the Oxford Transport Strategy (OTS) in 1999. This had the highly laudable aim of reducing air pollution, and so improve health and reduce damage to buildings, but without damaging the local economy. To achieve this, buses would replace cars for trips in and out of town.

The local transport planning authority is free neither to exert any significant influence over local rail services nor to build new rail infrastructure. For some strange reason, it *can* so intervene in bus provision.

The strategy has achieved little more than to exchange one polluting motor vehicle for another. The centre of Oxford has become no more pleasant as a result. It reeks of diesel exhaust and is just as noisy as before. Air pollution remains a threat to respiratory health, as well as to buildings. According to an academic monitoring project, business fell by ten percent on its introduction, and there has been death and injury among both pedestrians and cyclists.

Only bus companies have wrought any benefit.

Any claim that travellers have exchanged their car for a bus in Oxford is true only where they were left no alternative, *i.e.* where they were *forced* to do so. Any claim that the bus has "relieved congestion" is nonsense. Buses coagulate in the town centre, where they become concentrated.

FIGURE 2.4 Buses concentrated in Oxford city centre.

Oxfordshire County Council's Sustainable Transport Manager was reported admitting that buses are victims of the congestion it was hoped they would alleviate.[11] He failed to note that they also *cause* that congestion.

This is a known problem with bus service. The economist Anthony Downs commented that : [12]

> "... buses use the same streets as most private automotive vehicles ... As a result, bus travellers are often caught up in traffic-induced delays that plague drivers of private vehicles. Moreover, buses are relatively large vehicles that make frequent stops. Hence buses themselves are important contributors to general traffic congestion, especially in areas where many are concentrated together, as in large downtowns where many bus routes converge within a few blocks."

Reliance on bus services is often intended as a 'cheap' solution to congestion and pollution. We have seen that it does not appear to *be* a solution, and can now ask whether the cost we must meet is truly low, anyway.

Buses may be operated privately, but they are heavily subsidized. They fail to meet the cost of the road space they use, or even that of their termini.

In Britain, bus operators get a refund of between two thirds and four fifths of the tax levied on fuel. An operator pays less for their road-fund

11 *Cash Docked as Bus Use Slumps*, Oxford Times, 1st July 2005.

12 *Still Stuck in Traffic*, Brookings Institution Press, 1992.

licence than they would for a private car, despite the fact that a bus weighs anything up to eighteen tonnes, and thus incurs vastly more damage.

The problem of somehow estimating the true cost of building and maintaining roads on which buses can operate, is precisely the same as for heavy lorries. It has long been contended that lorries fail to pay for the roads they so quickly erode. Nearly half a century ago, US researchers reported that road damage varied as the fourth power of axle weight. This means that doubling the load would incur *sixteen times* the damage. Yet the road-fund taxation in Britain increases only around tenfold. For the bus, it merely doubles. True costs are met from general taxation and are rarely counted. One estimate puts this subsidy alone as high as £2,700 per year; say, around 5p/km, given an average annual bus 'mileage' of ~58,000kms. With the four-fifths rebate on fuel tax and the low fixed levy, the actual contribution made by bus companies is little more than 1p/km.

From the passenger's perspective, there is a huge variation in fares, according to region and journey length. Long-distance fares are extremely low, reflecting the greater speed and thus lower costs per kilometre. Current (2010) fares for return journeys into London of around a hundred kilometres (*e.g.* from Oxford) are around £16, *i.e.* just 8p/km. For travelling in and out of Oxford from a satellite town, twenty kilometres out, the figure rises to just over 12p/km. But more than twenty times as many bus trips are local, and of less than ten kilometres. Despite competition, local fares remain decidedly high, especially for one-off one-way journeys, where they can approach 60p/km. In Oxford, a return journey I regularly made of just three kilometres would cost me £2.20; nearly 37p/km (2005). This is very nearly the full cost of using a private car, with no other passenger, even though it neglects meeting the other five sixths of "track cost".

Admittedly, buses are cheaper in London, with an average 14p/km charged in the millennium year, where a strong central authority is able to apply a greater subsidy. However, increasing fuel and driver costs are causing fares to rise rapidly. Around Britain, in the five years to 2003, they grew by between twelve and *seventy-two* percent.

Guided bus-ways are the darling of a minority of transport planners, who see them as a "cost-effective solution", presumably to congestion. It is at least doubtful whether they are either a solution or cost-effective, especially in the longer term, and when existing rail infrastructure is underdeveloped or underused. The true intention is more likely to do something about the great weakness of the bus — its low speed. But by allowing buses to bypass congested areas, a bus-way removes the one advantage they can otherwise claim — many stops, close to the start and end of a journey.

A railway is not the consequence of some strange male obsession. The "permanent way" was conceived by hard-headed engineers as the cheapest and fastest way to carry the greatest load over the widest variation in terrain. The key ideas are to reduce rolling resistance by using rails, to raise occupancy by forming a train of vehicles, and to schedule a clear path, end-to-end.

A bus-way shares none of these features.

It may not even be safe. Steering is disabled and guidance typically achieved by little wheels that track the curb. Buses were never designed to be operated in this fashion. Safe operation is difficult to claim with so little experience. One system in Nancy had its licence revoked after just one month's operation, following two accidents associated with stability (one causing passenger injury). Safety is inherently lower with a bus in any case. Railway carriages are more expensive than buses primarily because of their very high safety standards: for example, they are designed to withstand a high-speed impact. Buses, on the other hand, are very lightly built.

Bus-ways are rarely independent of congestion. They typically deliver their vehicles onto a road shared with general traffic. There is no clear path end-to-end, as on a railway. Because carriages cannot be safely interconnected, capacity, as well as speed, is limited. In other words, a bus-way does nothing to relieve the limitations of a bus. Even the noise, stink, and air pollution are retained, right in the middle of town. In fact, these could be worse because the dedicated roadway would deliver vehicles at a greater rate, except that the town can only accommodate so many, and may already be saturated. Hence, any claim that the number of buses possible can increase is also ill-founded. Nor will the predictability of the bus improve. Congestion is still encountered; bunching and dispersion will still occur. Last, but by no means least: buses are wider than trams, and claim more space.

Given all this, it is very hard to see any case for the guided bus-way. The most common claim seems to be a lower cost of establishment, and the ability of the bus to reach the heart of the city. The latter argument cannot be sustained, on grounds of self-congestion and pollution, as we have seen. Any claim of lower cost has to be justified over a very long term, especially when new paths must be purchased and laid.

In Britain, a scheme in Cambridge is seeing a disused railway line destroyed and replaced by a concrete road. There is well-organized opposition who argue that true bus-way costs are being underestimated and hidden, while predicted passenger numbers are exaggerated. They propose reopening the pre-existing railway instead, which they claim will prove cheaper.

While true cost and ridership will remain unknown until something is up and running, the opposition make a strong case. A twin-track railway bed already exists, together with all engineering works along half the route. The proposed bus-way would require more land, all of which would have to be purchased and cleared. Being much wider, it would need new and longer bridges. The estimated cost of the bus-way has already risen significantly.

To highlight another objection to bus-way construction, I quote an authority, Paul Mees, with regard to a proposal for Brisbane, Australia : [13]

> "... Brisbane is set to commence construction of a billion-dollar network of radial express busways ... But ..., Brisbane already has an extensive, but under-utilised electrified radial rail network which the busway system will simply duplicate. ... One reason why the Brisbane rail system is so poorly patronised is the almost complete absence of feeder buses to rail stations. Only 7 per cent of rail passengers arrive by bus, because fares are not integrated and most bus routes parallel rail lines rather than serving stations. The busway system will actually formalise this situation; meanwhile, nothing is being done about multi-modal fares, timetable co-ordination between modes, or route redesign."

Far from each mode co-operating to create a holistic system (more than the sum of its parts), each mode competes with the other, to create a system that is actually *less* than the sum of its parts.

Modal integration clearly can be achieved, as it has been in London. Transport for London (TfL) found that, after walking, by far the majority of people arrive at a central light-rail station by either heavy rail or bus. The proportion using a bike is significant and rising noticeably.

13 *A Very Public Solution : Transport in the Dispersed City*, 2000.

There is now no need at all for bus and train operators to compete with each other. They surely have enough collective competition with the private car. One key to successful competition by shared transport with the private variety is surely effective modal integration, with the aim of reducing door-to-door journey time. Another is an effective information service.

Again, we can return to Transport for London for an example of what a competent authority can do. TfL offers an excellent internet "journey planner". One need only enter street names or postal codes and it will offer alternative routes, complete with an estimate of the time taken for each leg. Playing with this for just a few minutes makes clear the potential for co-operation between bus and train, in place of senseless competition.

Their recent addition of cycling has made its value abundantly clear.[14] The bike option is all too often far the fastest displayed.

A strong central authority over anything is sensibly regarded as an unpleasant, even dangerous, prospect — especially so when it applies to transport, which we have all come to regard as a private and personal affair. However, it would seem to be the price of modal integration. There is no reason why such an authority cannot be democratically accountable. Furthermore, there is no need for it to possess unbridled commercial power by acting as a universal provider. It need only plan, regulate, and co-ordinate, leaving others to invest, manage, and maintain.

Riding a bus is relatively safe, though published figures do not yet distinguish casualties on local journeys from those on direct inter-city services. In 2004, in Britain, buses accounted for 48 billion passenger-kilometres (bn pass-kms), divided almost equally between suburban and regional travel. (Although 98% of bus trips are local, non-local trips are much longer.) There were 20 deaths and 448 serious injuries, or 0.4 and 9.3/bn pass-kms, respectively.

While you may be safe *on* the bus, it poses a much more serious threat when you are on foot or a bike.

The bus is spelling the end of cycling in Oxford — a city once famous the world over for the image of the dusty don wobbling along on an ancient but sturdy mechanical steed.

---

14 An earlier draft was forced to lament its absence!

The need for a bus to repeatedly pull in and out across the path taken by a bike yields a conflict that is inevitable when they share a common conduit. Another threat arises when a bus turns a sharp corner, as at a junction. Being a long vehicle, its middle then rides over the inside edge of the road, where cyclists travel.

Bikes are the enemy-number-one of any bus driver. I speak from much personal experience. A conflict of interest breeds hostility and contempt, although simple neglect is often enough to kill. I have been on a bus when it collided with a cyclist. It knocked her clean off her bike and sent her sailing across the pavement. Her head missed a tree trunk by inches. She was lucky. In another incident, a bus driver was following a cyclist far too closely, as is commonplace. She fell off, and was run over and killed. The driver was convicted of manslaughter and sacked. He was chatting to a colleague at the time, but only needed a moment's distraction.

When mixing unprotected people on bicycles with large numbers of buses, serious accidents are inevitable. They also remove any perception of safety. No one on a bike can fail to be conscious of the risk with a ten-tonne bus following barely a metre behind.

There are only three safe options for any road: exclude the bus, exclude the bike, or provide a fully independent track for each one.

Car

Aside from any other appeal, direct access to a private car confers one undeniable and profound advantage over using the bus or train: a significantly lower door-to-door journey time is often possible. You don't have to divert to a station, wait for a scheduled service, and wait again while making a connection. Instead, you can leave when you like, and take the most direct route. For any trip of moderate length, this has an obvious appeal.

Precisely the same advantage follows with walking or cycling when travelling just a few kilometres. When going more than around fifty kilometres, it becomes worth a transfer to rail to benefit from the much higher speed along the greater part of the journey.

The car advantage depends upon one simple fact: most of our journeys are short. In Britain, the average trip length is just eleven kilometres.

Perhaps the best evidence of the importance of door-to-door journey time in our decision to use, or *not* to use, the car is an observation that inspired one Martin Mogridge.[15] He became fascinated by the fact that road travel in London always seemed to remain at a constant speed, regardless of changes in the volume of traffic or the space allocated. The only explanation seemed to be that once congestion slowed your car beyond a certain critical point, it became quicker to make the journey some other way. In London, for example, there is an excellent light-rail network.

A corollary to a proper understanding of this is that the best way to make road travel faster is actually to invest in the competition — rather than improve the road system itself, which might just induce new traffic under *Downs' Law* (see next section), we improve the alternative. Make it possible to complete a journey by bike and/or rail in less time and more people may "shift mode" from the car.

Lighter traffic on the road will then mean a higher speed all round.

In both London and New York, around half the population have never bought a car because there is usually a quicker way to travel. A high population density, with strongly focal distribution, has justified sustained investment in shared transport infrastructure.

Elsewhere, there is typically a lot less choice.

Besides genuine transport utility, the car serves as a primary affectation. While we naturally dislike discussing such things, only a fool would deny our need for them. A world where no one expresses their individuality and taste is not one we would wish to even visit, never mind inhabit.

To quote Aldous Huxley : [16]

> "In the course of evolution nature has gone to endless trouble to see that every individual is unlike every other individual. ... Any culture which, in the interests of efficiency or in the name of some political or religious

15 Martin Mogridge was a leading transport consultant and Professor at the Centre for Transport Studies at University College, London, from the late 1970s on. He is important in bringing mathematical rigour from his training in theoretical physics to the analysis of transport. His taste for brightly coloured skin-tight leather suits was equally unusual among both academics and transport engineers. Sadly, he died young in 2000.

16 *Brave New World Revisited*, 1958, Chapter 3.

> dogma, seeks to standardize the human individual, commits an outrage against man's biological nature."

Deep down, we are instinctively aware of our individuality and its importance. Threats to it can expect resistance. Opportunities to express, and maybe exaggerate, it are seldom lost.

The car arrived at the end of a period when people, particularly men, had seen their individuality compromised as never before. During the industrial revolution, they had been forcibly reduced to mere components within a machine, and equally replaceable. Through two terrible wars, they had marched in uniform to the beat of a single drum. No other generation would perhaps have welcomed the arrival of the private car as much as the one walking by the showrooms in the 1950s and 1960s.

Of course, there are other ways to express individuality. Learning this is part of growing up. Excessive reliance on adornment, of any kind, is typically juvenile. I found myself surrounded just the other day, on the train, by youngsters with green waxed hair, black make-up, and enough facial jewellery to fund a pension. We can all be forgiven for such excess in our adolescence, and for hanging on to what became precious at that time: from styles of clothes, through music, even to philosophy and politics. Such things often grow into luxury and indulgence, as with the car.

They remind us of our youth.

About the most powerful image, in this light, that I can recall is a TV article about a yellow Lamborghini — a thing of great beauty that made all the right noises. As it screamed to a halt, another thing of beauty stepped gracefully out, to the music of James Brown. To those of us who recall both the music and our own cars, more humble though they were, it brought a tear to the eye. No one ever fully admits leaving those things behind.

But then we all have to grow up.

Not everyone can enjoy a car as an option in the first place. To do so, you must first be able to afford one, and be fit and qualified to drive. Fitness is a problem for the old, and the cost of gaining a licence one for the young. You must then find the space to keep your car at home.

A majority in the rich world have managed all this comfortably for decades, but that world is now changing because of its very prosperity. The sheer *number* of automobiles has come to threaten their utility. For example, students at Oxford University are disallowed the privilege of keeping a car for a very simple reason: there is just not enough parking space in a mediæval city to accommodate the high proportion who would bring them.

When the car was the fancy of a fortunate few, parking was no more an issue than was congestion. It is now a very different matter. Lucky are they who can park outside their office door. Long is the queue at the school gate at pick-up time. Expensive is the home with adequate off-street space.

Parking is the bane of modern life.

California is an odd place to someone who grew up in a world built in Victorian times. I recall attempting to visit a 'diner', across the road from my hotel, on foot. There was neither 'sidewalk' nor crossing. When I walked and crossed anyway, horns were blown and abuse hurled. It was as though, as a pedestrian, *I* threatened *them.*

Wondering around San José, I saw car parks the size of towns. When you parked, you would find the building on the horizon. There was no option but to run buses to bring you to the door — an apparent admission of defeat. Only in Los Angeles can you find a more profound contradiction. Satellite images indicate that around 43% of the territory is under tarmac, yet I was to experience the slowest mechanized transport of my life. We gave up with the jams and left the area at the first opportunity.

Los Angeles is proof that road transport alone simply doesn't work.

Many are the valiant who continue to try to bring forth alternatives, but they are up against it. The place was built for the car — or rather *re*built. Like San Francisco, up the coast, Los Angeles once ran trams. Unlike its sister in the sun, LA ripped up or buried the rails. An attempt to ease congestion by building new lines has been met with only partial success. The traffic heading for a station can itself cause a jam, and there is a tendency for *any* additional transport capacity to encourage yet more travel, anyway.

Sprawl is a cul-de-sac out of which you cannot easily reverse.

Infirmity and disability become more common with age. My mother never learned to drive. There was neither the need nor the money to do so while she was young. Her mobility became limited to shared transport that either remained commercially viable or has been maintained through subsidy. My father voluntarily surrendered his licence when he became aware of tunnel vision, which made it impossible for him to maintain a steady heading when looking to the rear or side of his vehicle. Since fitness to drive is not regularly assessed, we must presume there are many on the road with this lethal condition, especially among those in the later phase of life.

While ill health may come to exclude many nearer the end of life from driving, the largest group to be denied their independence is the young.

From the age of six, I would take myself to school. Initially this meant walking a couple of hundred metres. From age eight, it meant a one-kilometre walk or bus ride. From ten, I would walk the five hundred metres to the station and ride around ten kilometres by train, needing a single change halfway. Closure of the line along the first half left me with a journey by bus that took twice as long and cost twice as much. From that point on, Dad would take me in the car, on his way to work.

The "school run" now accounts for a *quarter* of morning traffic.

One of my greatest joys is in watching my children learn. It is absolutely clear to me that their instinct for play is directed at learning to do the things we grown-ups do around them. It is equally clear that our part of the bargain is to harden our hearts and wean — to let go the hand and let them set sail on their own. Nothing concentrates the mind more than coping on your own. Nothing connects you to the world more firmly.

And yet we now deny our young *any* opportunity to set forth alone.

In the aftermath of the July 2005 bombing of London, certain streets were closed to through traffic. Children briefly emerged out onto the streets to play, as their parents once had. Soon the stripy tape came down; back came the cars; back indoors to their almost solitary confinement went the kids.

I had seen something similar at home. Many years ago, our street closed to traffic as a result of road-works outside the village. Everyone drifted slowly out onto the street and we discovered neighbours we never knew we had. Children suddenly appeared. Until then, I had been unaware some of them even existed. A snapshot followed of a once-cherished liberty.

You need to see it to understand.

No museum could credibly recreate the sight.

The only reason the number killed and seriously injured in the street has fallen to a level we seem to find communally acceptable is that children have been removed from the threat, rather than the threat from the children.

Children depend upon walking and cycling to get around. As a pedestrian, you need to cross the road frequently, often in situations that make it dangerous. On-street parking can often make it difficult, sometimes impossible, to see what is coming. My mother lived on the inside edge of a curve on a road that had become very busy. Parked cars rendered it impossible to see anything like far enough. At eighty, with two hips replaced, she was no sprinter. Every day she played "Russian roulette". She had no choice.

A pedestrian crossing, we were told, was unaffordable.

Children find themselves in a much worse position. They cannot see over the tops of cars, which have in any case become higher over the last few years. Until aged around ten, they are physically unable to judge speed and distance adequately. They are thus reduced to dependency. The threat is considerably increased by the way in which they now arrive at school. Every morning, parents fight for the inevitably limited space in the lay-by in front of our local primary. On the pavement, children gather excitedly, running about as their parents meet and organize.

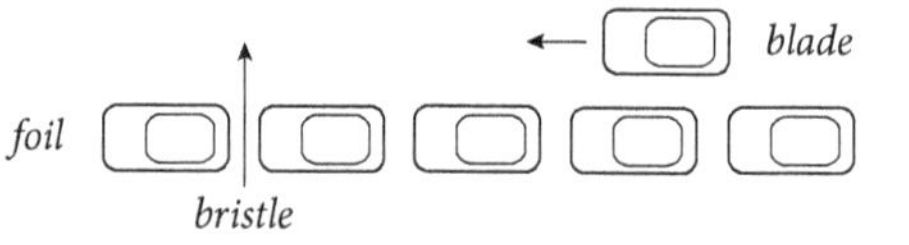

Figure 2.5 A *razor* formed by parked cars.

A *razor* has been formed right next to them, with the row of parked cars as foil. All that is necessary is to add the blade in the form of a passing vehicle. It need only be moving at 30kph. That will be enough to cut down a child as they dash innocently through the foil, like a bristle on their father's face that morning.

I have seen it happen. Thankfully, after some burned rubber, steel did not meet a little skull that time.

Each year, in Britain, around *thirty-three thousand* pedestrians suffer injury, and about seven hundred death, by a road vehicle. For every four casualties among users of the road, one is a pedestrian. About one in three serious injuries are inflicted upon children, and about one in eight deaths. Children are very much more likely to suffer serious injury than an adult, when struck by a car. Their heads are lower and softer, and they are much more likely to be drawn beneath the vehicle. The recent fashion for larger, higher, cars, increases the threat. "Bull bars" are simply lethal.

Around seventeen thousand cyclists suffer serious injury each year by road vehicles, and nearly a hundred and fifty death. Children on bicycles have all but disappeared. It has become more or less pointless them owning one. Many, if not most, parents now regard cycling as simply too dangerous for children. The threat arises entirely from motor traffic.

Much literature is devoted to such problems in the urban environment, but exactly the same injustice and lost liberty is suffered in the rural one. Just like the urban street, the country lane has been commandeered for the exclusive use of the car. This was never the subject of public debate. Country lanes vary continuously, and almost invisibly, in width, making cyclists vulnerable to ever more and wider cars, driven at ever greater speed.

In both town and country, as noted before, the two most intrinsically safe modes of transport — walking and cycling — on which the young in particular depend, have been rendered by far the most dangerous.

Believe it or not, the vast majority of the road network was neither designed nor built to serve the car. The car was to take it away from those for whom it *had* been constructed. The reason it was able to do this was simple. Early motorists were, by definition, rich, and, in consequence, influential. Few indeed could afford the luxury of a motorized private carriage in 1900.

At that time, the road had never before been as peaceful or unpolluted. After the arrival of the railway, horse and carriage were now favoured only for the most local of journeys. This also had the effect of removing obstacles along the road. Turnpikes went broke. Toll roads and bridges passed cheaply into the hands of parish and public. In town and country, the road had become the province of walker and children at play.

By the 1880s, the road network was undoubtedly underutilized. With the arrival of the bicycle, complete with its new pneumatic tyres, this was to change. For a while, one could cycle safely and in peace.

While principal roads had been massively improved by the process of John Loudon McAdam, the majority of local roads remained in a sorry state. Cyclists were to complain bitterly and campaign strongly, but to little effect. These were ordinary folk with ordinary means. They were ignored by the ruling élite. The motor car lobby, beginning in 1895 with the wonderfully named *Self-Propelled Traffic Association*, were a very different kettle of fish, and were to gain everything they asked for. They were limited to just 4mph in the country, and 2mph in the town, and had to be preceded by a walker with a red flag by day, and a red lantern by night. By 1896, the speed limit for "locomotives on highways" had been raised to 14mph, which would have made it rather difficult for the chap with the red flag to stay ahead.

Early motorists — "cads on casters" — were far from popular, causing widespread damage to mostly quite inappropriate roads. They also wreaked widespread injury and death on both man and beast. Even in those days, the response of many a driver, on being caught, prosecuted, and fined, was to complain of 'persecution'. Speed limits were vigorously opposed, and widely ignored, as they are today. The car was claimed to be a necessity for productive employment and trade, rather than a luxury — somewhat debatable when there were barely twenty thousand in a nation of forty million. In truth, it was then a toy of the rich used almost entirely for pleasure.

Despite the dire and numerous complaints of ordinary folk, and of local authorities who had to meet the cost of repairing the damage done to the road, Government was to offer even more than the motorist demanded. When it did finally raise a tax on motoring, it was 'ring-fenced' to improve the road by finally macadamizing everywhere. Though this did placate local authorities by paying for repairs, and many among the public by removing the "dust menace", it was done overwhelmingly for the benefit of the car-owning class, which just happened to include most members of parliament.

Ironically, the car lobby opposed the Government proposal to build new roads, for the exclusive use of the car and free of any speed limit. They were canny enough to foresee their undoing. It would raise their taxes, and make it harder to remove restrictions, including the speed limit, on ordinary roads. They managed to appear conciliatory, requesting no special treatment, when what they actually wanted was nothing less than unrestricted use of the entire existing road network.

Which would have the consequence of excluding everyone else.

Before the car, people still died on the road, and in surprising numbers. Certainly, there were fewer people around, but then there were fewer roads, and many more goods had to be transported to and from the railway depots along them. The very freedom children then enjoyed placed them in some danger, though nothing to compare with what was to drive their descendents indoors. In 1875, for example, before the car arrived, there were 1,589 deaths on British roads. By 1909, around fifty thousand cars were using them, and were 'involved' in 373 fatal accidents. Within another five years, cars were killing a thousand more each year.

A favourite cry of the car lobby is that speed has little to do with accident and casualty rates. This is unsustainable in the face of the facts. When they succeeded in gaining abolition of the 30mph national limit in 1930, the casualty rate rose by a quarter, from 171,000 in 1928 to 213,000 in 1932. They would doubtless point out that nearly two thirds of road casualties today occur *within* a 30mph limit. To a fool, this might suggest that we would be better off without a speed limit. In fact, it strongly suggests the reverse — that action is needed to reduce speed further in these locations, to prevent or reduce injury. About a third of all fatal accidents are caused by speeding, and it remains a factor in many more. But in *any* accident, vehicle speed will exacerbate the effect of collision, increasing injury and the risk of death. If all accidents occurred at speeds below 10mph, few indeed would die.

Nothing is more hotly contested by the road lobby than the need for a speed limit — the last remaining restriction upon the car on the public road. They appear to believe that it is an unnecessary interference with personal freedom. A little physics suggests otherwise. When speed doubles, the kinetic energy of the vehicle *quadruples*. To reduce speed, brakes apply a constant force acting against motion. The period of application varies in direct proportion with kinetic energy, not speed. Hence, twice the speed means *four times* the braking distance.

Should you hit someone at greater speed, more energy will be transferred to bones and into vital organs. Our bodies are, in fact, designed to sustain shock — to absorb a certain amount of energy — but only so much. Imagine playing golf with an egg. If you use a wooden club, and tap it gently, the shell will absorb the shock and the egg will bounce away. Use a metal club, and hit it hard, and it won't move at all. It will just make a mess of your iron. It's essentially the same with a child's head. This is why car manufacturers are being compelled to make softer bonnets, and why even a modest excess in speed can indeed kill.

An Australian academic study, funded by the Federal Office of Road Safety, found that, on a road subject to a 60kph limit, the risk of becoming a casualty in a crash doubles with each 5kph increase above that limit.

Physiology, physics, and the bloody aftermath of collision, make clear that existing speed limits remain too high or are inadequately enforced.

Speed kills, period.

In a world where unavoidable disease is giving way to disease brought about by irresponsibility, such as abusing alcohol and eating junk food, it has been estimated that road accidents may become the cause of the third biggest loss of so-called "disability-adjusted life years" (DALYs) by 2020.

With the car, it is not enough to pass judgement as to its safety in conveying you. One must add the cost to innocent bystanders.

In 2004, 3,221 people died on British roads and 31,130 were seriously injured. While the reported death rate may be reliable, the account of injury is not. The Department for Transport (DfT) considers that the true figure may be underestimated by at least a factor 2.7. But there is more.

Another error lies in relating casualty count to distance travelled; in particular, in accounting properly for vehicle occupancy. If fewer people were actually making a journey, then fewer kilometres were actually being travelled, and the casualty rate *per passenger-kilometre* understated. The difficulty lies in distinguishing just how many passengers there are.

When two people are sighted occupying a car, it does not imply both wished, or needed, to make that journey, any more than when two are counted on board a bus. One will then be the driver — there only to *facilitate* that journey, in the same way as the engine and wheels. Perhaps the most common example is a parent taking their child to school. There is but one passenger; the parent is merely the chauffeur.

As options for independent travel disappear for the young and infirm, the proportion of 'taxi' trips is rising. Official figures take no proper account of these and thus overestimate car passenger occupancy. The DfT reports an average occupancy of 1.6 passengers per car. While the average *vehicle* occupancy may be 1.6, published statistics allow the figure for vehicle *passenger* occupancy to be reduced to about 1.2.

The official estimate of car occupancy is derived from a survey that returned, for example, in 2003, an average distance travelled of 5,238kms as driver, and 3,165kms as passenger. There are two extreme interpretations; that all accompanied trips carry just one passenger, plus chauffeur, or that they all carry two passengers. (Trips with more than two occupants are sufficiently rare to safely ignore.) In the first case, passenger occupancy would always be 1.0; in the second, it would be 1.6. Figures for the proportion of trips made by purpose, such as the school run, allow an estimate of the truth between extremes.

Almost 500bn vehicle-kilometres were recorded in 2004. To calculate the casualty rate, we must multiply this figure by the *passenger* occupancy — roughly 1.2 — to yield about 600bn pass-kms. It seems reasonable to assume that everyone within a vehicle, of any kind, shares the same risk, so we retain each casualty figure, unaltered. If our accounting is correct then there were around 5 deaths and 52 serious injuries per billion passenger-kilometres.

Recall that these figures reflect casualties both suffered *and incurred* in the cause of travel. Only half those killed on the road were car drivers or passengers. A full quarter were those engaging in walking and cycling, each of which, as we have noted previously, is otherwise almost entirely safe. We must compare the 5 killed in their car with the 39 cyclists and 52 pedestrians, ... and forgive the non-driver who gives up and buys a car.

Cycling has declined *sixfold* since 1952, precisely mirroring the rise of the automobile. Nearly 7,400 pedestrians were killed or seriously injured in Britain in 2004, a third of them children. Even rail passengers are not secure against this evil. Of the twenty-one rail fatalities in the first six years of this millennium, no less than fifteen were caused by car drivers.

To drive exacts a price in the safety, and thus liberty, of others.

An internal combustion engine (ICE) is well known to liberate toxic fumes, given the hydrocarbon fuel it burns, wherever it runs. This issue will be fully considered in Chapter 4. For now, we shall be content to stick a marker in the sand. ICE air pollution causes *serious* disease, including cancer. A continuous effect upon all of us is a veritable plague of 'minor' ailments.

I recently stood on a bridge near my home over a four-lane highway. (It forms an ideal vantage-point from which to photograph a congested road running right alongside an underused railway.) After just twenty minutes exposure, I found myself suffering several hours of tension, irritability, fatigue, and a most strange mental state I cannot begin to describe. Symptoms such as this, often accompanied by severe headaches, are documented effects of ICE pollution. Having suffered most of them, I can only wonder at the silent suffering of anyone living on or near a major road.

Urban areas, where traffic is concentrated, cause huge plumes of pollution that can fall out many miles away, over innocent and guilty alike. So it is no use thinking you are safe as a country bumpkin. We are all at risk. In fact, we are probably all affected, without making the association. Some of us will suffer more than others, either because we are more exposed, without perhaps knowing it, or because we are more susceptible.

The worst place of all to be is *inside* a car, in traffic.

Other problems arise from the dependence of the automobile on an ICE. These also are fully explored in Chapter 4, but a start here is useful.

Every ICE emits carbon dioxide that adds to the greenhouse effect to cause climate change. It is also inherently inefficient in its consumption of energy in two different ways. First, it can convert at most 70% of the chemical energy in its fuel to work done by a rotating shaft. Second, its power delivery drops off dramatically as revs decline from a narrow optimal band. This is why you need a gearbox and must "rev up" to pull away at the lights. The efficiency of the entire 'power-train' typically falls below 25%. This is further reduced by a factor ten (or more) according to the "mass ratio".

To add insult to injury, almost every vehicle currently on the road repeatedly throws away the kinetic energy so painfully gained. Only a handful of 'hybrid' cars employ *regenerative braking*, where much of this energy is recaptured. There are many ways to accomplish this trick, and the pay-off is considerable. The most common method is electrical.

It would be easy to imagine a conspiracy to withhold regeneration. Think of all that additional fuel that can be sold, and how vehicle production costs can be kept down. The usual failure of imagination is perhaps more likely, but ties between oil companies and car manufacturers are not unknown.

There is one more damning indictment of the car with regard to energy waste. Any automobile, by definition, must carry its own energy supply (*i.e.* fuel) with it. An 'off-board' energy supply avoids the need to propel the extra mass implied, and allows for greater efficiency in generation, even after losses in distribution. And you can't regenerate fuel.

This is why electrification is such an advantage for a railway.

As we established in the last section, by far the most important cost considered when deciding how to travel is the time taken to complete the journey. For people who charge by the hour, this cost is immediately perceived in financial terms. For the rest of us, it is perhaps measured instead as time away from our family. This is why we curse 'traffic' so much, usually disregarding the fact that we ourselves have added to it.

Traffic jams arise because, as more vehicles join a conduit, the space available to each one diminishes. As the gap separating them reduces, their speed must fall in order to maintain a safe braking distance.

When setting out to calculate a safe interval between vehicles, there are three things to consider. First, and simplest, is the vehicle length. If your vehicle and the one in front are each twenty metres long, you will have a little trouble keeping just ten metres apart. If you do manage it, it will probably involve a crunching sound, and prove expensive, if not painful. The minimum separation possible, measured centre-to-centre, is the same as the length of the vehicles concerned.

A second consideration is reaction time. You need a period in which to perceive the need to stop, and then apply the brakes. In that time, you travel a distance proportional to your speed. If you move twice as fast, you would go twice as far. You need a separation that varies *linearly* with speed to avoid a collision.

The final element is the distance you need in which to bleed off kinetic energy. Everything that moves is said to possess energy, by which we mean you have to apply a force and perform work to stop it. Physics (or any other science) begins with observation. Anyone who investigates will see that twice the mass needs either twice the force, to stop in the same distance, or twice the distance with the same force, to bring it to a halt. This is everyday experience, but it took Isaac Newton to recognize it as a 'Law' of nature — something we must simply accept — and explore further.

You might think that the stopping distance for a thirty-ton lorry would be a tad more than that for a typical family car at the same speed. Provided the braking and traction systems are properly designed, to deliver the maximum braking force possible, without slippage, and the road surface is the same in both cases, the distance required to bring each to a halt will, in fact, be the same. While the kinetic energy is greater for the lorry, so is the maximum braking force. The largest back-force that can be achieved via friction, before the wheel begins to slide, varies in proportion with the weight acting on it, and therefore with vehicle mass. One effect cancels out the other. True, more braking force is required, because more mass means more kinetic energy, but more force is available.

Newton also noted that if the same mass was moving twice as fast, stopping required the same force applied over *four times* the distance. While greater vehicle mass need be no problem (except for the engineer designing the braking system), speed is decidedly different. Moving vehicles twice as fast along a path would indeed help to increase the flow of people, but is more than counterbalanced by the increased separation necessary.

Doubling speed actually *halves* the flow (Figure 2.6).

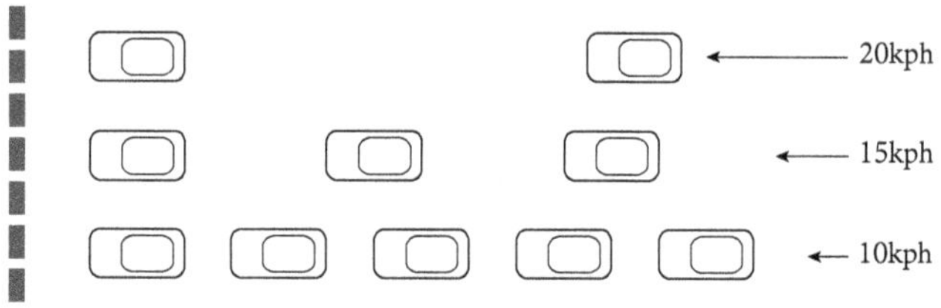

FIGURE 2.6 Effect of doubling the speed on uncoupled vehicles.

As speed increases, the rate at which vehicles, and their occupants, pass over a fixed mark in the road decreases because the interval between them must grow disproportionately. Conversely, each time an additional vehicle joins the procession, it invites a smaller interval, simply because the same space must be shared by more vehicles (Figure 2.7). This, in turn, compels *everyone* to slow down, and thus to suffer an increase in journey time.

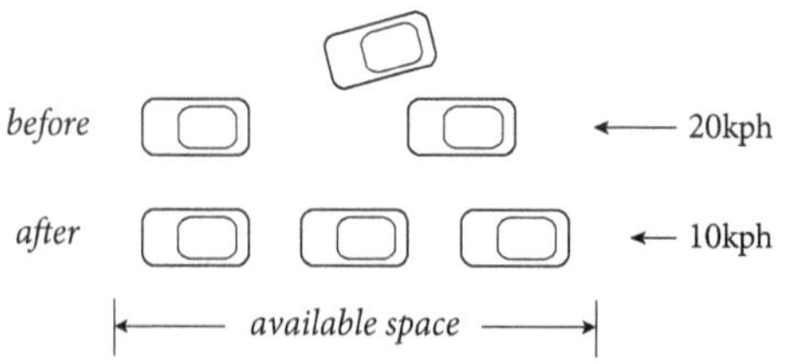

FIGURE 2.7 Effect of adding one more vehicle.

An alternative is to couple vehicles to form a train (Figure 2.8). By adding carriages we raise *occupancy*, allowing more passengers to pass the mark per hour. The safe separation we previously required between carriages need only be kept between entire trains. Speed can then remain the same.

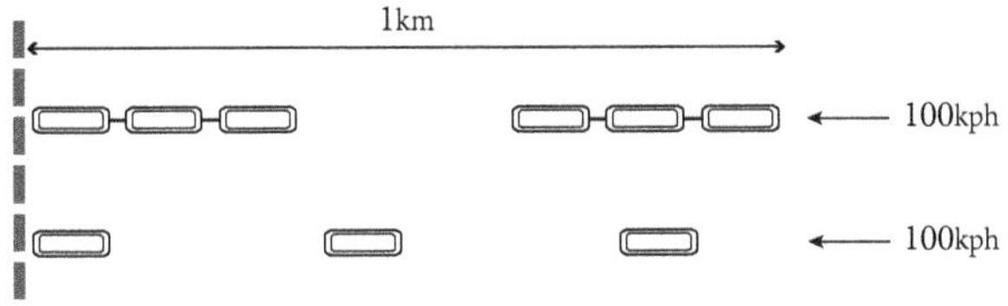

FIGURE 2.8 Coupling vehicles increases flow.

Another way of looking at it is to consider the occupancy of each carriage in isolation, which might be the same as when uncoupled. While speed can also remain unaltered, the *frequency* with which vehicles can be placed on each kilometre of track (road) is free to rise.

Flow is calculated by multiplying together occupancy, frequency, and speed — people per vehicle, vehicles per kilometre, and kilometres per hour (see next section). By raising either occupancy or frequency, according to your preference, forming a train increases the *capacity* (maximum flow) of any available conduit. It therefore increases the value of that conduit.

An additional advantage is that we no longer need either a power unit (engine) or a driver in every carriage. Vehicles become cheaper per seat, and a great deal of weight is saved. Lighter carriages need less propulsion.

Much has been written about the possibility of coupling together automobiles to form a "road train". No credible system has ever been demonstrated. The difficulties involved in establishing a system that lacks the mechanical guidance of a railway are immense. It may never be possible to guarantee the necessary safety. Any system capable of dynamically forming and breaking up, as has been proposed, would necessarily be even harder to render safe, and would retain an engine, at least, in each carriage.

In any case, much of the appeal of the car is its independence. A road train would appear to deny the principal advantage of both rail *and* road.

With the railway, in Britain since 1993, private vehicles run on public track. Despite a widespread perception to the contrary, this is *exactly* the case with road. Private vehicles run on public roads. In common with every other nation, the highway is communally owned and maintained.

As mentioned earlier, the car lobby had a long hard fight to gain access to the public road network, and won largely because early motorists were by definition rich and influential. Motoring was a leisure pursuit of the gentry. It thus quickly attracted the attention of the tax raisers. Asquith[17], Chancellor of the Exchequer, noted in 1907 that a tax on cars would be :

> "... almost an ideal tax, because it is a tax on a luxury which is apt to degenerate into a nuisance."

A proportion of this tax would go to local authorities, who had responsibility for roads — not just for repairs, but for their improvement. The public road network was thus surrendered to the motorist and road-freight operator, who would henceforward claim ownership. However, unlike rail operators, they had *not* paid for its construction. Nor had they an exclusive claim.

At today's prices, construction of Britain's 3,500kms of motorway and 8,000kms of dual carriageway alone would cost around £110bn. Then there is around 38,000kms of principal, and over 300,000kms of secondary, roads. Were it new, the total value would easily exceed £500bn. Even when reasonably written down, we speak of an all but priceless communal asset.

UK road-use taxes are charged in the form of fuel 'duty', raising £22.6bn in 2004, and a "vehicle excise duty" (VED), raising another £4.9bn. Of that, Government and local authorities spent £7.5bn on maintenance.

To this day, it remains utterly unclear to what extent drivers pay for the roads they use. Certainly, tax revenue exceeds expenditure, but we all accept that the State must raise revenue where it can to meet the costs of all the services it provides. There is also the question as to what extent the road network can be regarded as 'heritage' — a gift from our ancestors — whose cost can thus be written off entirely. This notion was obviously unfair to railway operators between the wars, who continued their obligation to pay dividends to those who shared the costs of construction. To charge nothing directly for access to the road denies fair competition with rail and leaves no mechanism for fair treatment among all users, large and small.

Maintenance is *not* the only cost associated with road use.

Like the railway, many roads require signalling. The policeman directing traffic at a junction is now the stuff of old movies, but, in fact, there are now many more police resources tied up with managing the road. Motorways and

---

17 Herbert Asquith was later to become a long-serving Liberal Prime Minister and famous for introducing health and unemployment insurance for the entire working population, and for ending the right of the unelected House of Lords to deny legislation, in 1911.

urban highways are monitored via closed-circuit television (CCTV). Traffic levels are monitored and congestion reported. Drivers frequently require assistance. Criminal behaviour must be dealt with, though driving is now subject to far less policing per vehicle-kilometre than ever before, so that dangerous practices like speeding and tailgating have become epidemic. The level of policing funded is clearly far from adequate, but the cost of doing the job properly and restoring order remains unknown.

Then there is what used to be known as the Fire Service, now renamed the Fire and Rescue Service because it devotes such a large proportion of its time to road accidents. The Ambulance Service and hospital Accident & Emergency departments have become similarly burdened. All are maintained through general taxation. Since 1998, in Britain, motoring insurance that pays compensation for injury or death in an accident has been required to pay also for hospital treatment, up to a limit of about £3,000. It seems unlikely that this covers the full aftermath of a road accident. Other costs include long-term hospital care, legal processes, and lost production. One estimate put the true annual cost as high as £17bn (2005).

This leaves the taxman with rather less than we might have thought.

So far, we have only mentioned costs that could be counted fairly easily, should an administration decide so to do. Other costs are less easy to quantify. The UK Royal Commission on Environmental Pollution (RCEP) suggested the following additional annual 'external' costs (1994) :

— congestion	£15–20bn
— air pollution	£2–5bn
— other pollution	£8.3bn+
— climate change	£1.5–3.1bn.

All of these must be met either in taxation or lost production.

Congestion is notably by far the biggest of these "difficult to quantify" costs. In case £20bn sounds rather harsh, consider that there are currently (2005) twenty-six million cars licensed in Britain. If each of these are delayed just fifteen minutes each working day of the year (a rather conservative estimate), and the average full employment cost of a single occupant is just £10 per hour (again, conservative), then a reasonable estimate of lost production comes to £15bn — the lower bound on the RCEP estimate.

The Confederation of British Industry (CBI), not famous for its green politics, recently concurred with the *upper* bound — £20bn.

Free parking, both in town and at our place of work — the most common destinations — is slipping into history. One leading corporation recently estimated the capital cost of providing an employee parking space at more than £800, plus £110pa maintenance. Multi-storey car parks have been estimated at nearer £6,000 per space, with correspondingly higher maintenance costs. Parking must compete with other claims on land.

There are also aesthetic, environmental, and safety, concerns.

Something that has long been enjoyed for free is now becoming a burden on both the public and private purse. Both "city hall" and employer are moving rapidly towards charging the full market (and external) cost.

Even the true cost of parking at home must now be faced.

Parking charges contribute to the *operating* cost of a private car. Of course, this depends somewhat on the nature of the vehicle. We shall sadly forego the large Italian racing thoroughbred for something more typical, though less alluring perhaps to the eye and ear — the ordinary family car.

One way to learn the true cost of running your car is to give it up.

In 1996, I left academia and strode out to pay the mortgage through private enterprise. I sold my services as a computer systems consultant and software contractor. Mostly, this was great fun, and paid twice what I was used to receiving. I had to travel about the country to meet and talk to clients, which proved most conveniently done by rail. Even better, I was able to work *en route*, meaning I was actually earning while travelling.

After a few months, I realized that the car had not been out of the garage, but for the odd trip to visit family and friends. So I sold it.

My finances saw *two* revolutions: far more came in, far less went out.

After three years, I felt again the call to learn and teach, and returned to students, research, and academic life. I deliberately opted for a post that would allow me to retain an empty garage, and all the money a car takes to run. My commute was a thirteen-minute train ride, plus a fifteen-minute cycle, each way. I found this pleasant, social, cheap, and healthy. For one full year the good life lasted. Then came the news that the department was to move to an out-of-town site. At seven miles, it lay no further from my home, but along a lane made too dangerous for cycling by varying width and unpoliced driving. There was no direct bus or rail alternative. A seven-mile commute by public transport would now take an hour and a half on a good day. Like countless others, I was *forced* to become a 'motorist'.

The car we bought is a typical family vehicle. Our use of it is similarly typical. One feature is particularly useful. A dashboard system computes and displays fuel consumption, both immediate and average. If you don't have such a thing, get one. I have found it easy to train myself to reduce the fuel I use by at least ten percent. As a result, I regularly record an average consumption of forty miles per gallon (14kms/l). In 2005, the price of gasoline in Britain was 91p per litre, giving a fuel contribution to running costs of 6.4p/km. This is not a typical figure. For one thing, I drive for economy, not the kick in the back. Second, I chose a fairly economical car. The national average gasoline consumption is just *thirty* miles per gallon, and diesel forty. Overall, the average is thirty-two, yielding 8p/km.

Transport in general suffers the peculiarity of high fixed costs. Planes, trains, and automobiles, cost a fortune to manufacture. Only the smaller proportion of their cost is incurred by actually using them, this being fuel and the additional maintenance needed. In five years, the only *running* maintenance we have applied to our car came when we were compelled to replace its tyres; one had been punctured by a nail, the other three were worn out anyway. Running maintenance thus added just 1.5p/km.

Fixed maintenance is a different story, and contributes to the cost of *ownership*. The annual service charge rises rapidly with each passing year. Our fourth annual service cost £430, to which must be added £60 for a new battery, replacing one that developed a fault. As this annual cost rises, it is offset by falling depreciation, which has now reduced to around £1,000pa. A simple way to calculate depreciation is to just divide the renewal cost by life expectancy. Saving that much each year will produce enough cash to replace the car when it has become worthless. Price inflation can be balanced against interest earned. We will need £18,000 in fifteen years, at today's prices, requiring £1,200 to be saved each year.

The fifth year in a vehicle's life is not atypical. Certainly, costs will fall overall as it grows older still, but only to a modest degree. Even this can be offset by the loss of production or leisure caused by dealing with the rising frequency of fault and failure. I would not recommend running a *very* old vehicle. My life has come close to forfeit at least twice in old cars — brake failure on both occasions. I would never take the risk again.

Roadside assistance is nowadays essential; a modern car is beyond "doing it yourself". The easiest way to provide for this is via an insurance policy. If you breakdown, all you have then to do is call a number and someone will come... eventually. Assistance should include transport to home or destination, of both family and vehicle. A typical policy, effective within Britain, costs around £80pa.

Next comes road-use tax and insurance. In Britain, we pay the aforementioned VED in return for a licence to keep and run a car on the public road. For our car, this is currently (2005) £165pa. Our insurance is "fully comprehensive". Any lesser policy would just infer a degree of self-insurance, that would justify its own annual budget. Renewal this year cost £270, with the benefit of 60% "full, no claims" discount. This is below the average paid, but we shall again err on the conservative side.

The incentive to use a car, once bought is considerable. After all, our annual cost of ownership of £2,205 will remain the same whether we drive 3,000kms, as we do, or the national household average of 20,000kms.

Not all cars are used for commuting, especially now that rising incomes and falling prices mean that we no longer share a single car, even with our partner. Many households now run two cars. The second is used less, reducing the average 'mileage' for all cars in Britain to just 14,500kms (2004).

Primary ownership costs thus contribute an average 15.2p per kilometre driven. Despite the predictability of every contribution to this figure, most people seem to consider only the fuel cost when using their car.

While US fuel tax is just 16%, compared with 68% in Britain (down over the last few years from 75%) there is no evidence of any consequent difference in behaviour. If we Brits had the same level of taxation, we would pay just 33p for a litre of 'gas'. Undoubtedly, this would produce an initial "drive for joy" but it might last no longer than the 'shock' that invariably follows each price rise. The fact is that dividing the fuel price by three would shave barely a *fifth* off the cost of driving we have counted thus far.

In reality, the effect will be even less because, at around 25p/km already, we are far from done. There are several other contributions to the cost of motoring that are very rarely taken into account.

You might be one of the lucky few who do *not* have to borrow money to buy your car, but you would be wrong to celebrate. Even if you have the necessary cash in the bank, you will still face the loss of interest the bank would have paid *you*. At the 5% annual-equivalent rate (AER) that can be easily found, as I write, that sum would have earned you £750pa.

That adds another 5p/km, with an average mileage.[18]

It will be more if you need to borrow. The essence of banking is to charge more for loans than is paid for savings. They keep the difference and are thus keen to keep that as large as possible. Nor are they the only ones to make a buck from your vulnerability. Car vendors usually make significantly more money selling finance than they do selling cars. Here and now, with collateral, you might get away paying an AER of 6.5%, or 6.5p/km. Without collateral, the rate will be higher.

Reflecting the high proportion of car purchases that are financed by a loan, we shall assume an (again conservative) average 6p/km.

Note that you do not escape even when you pay off the loan before the car is worn out. You will still be left with equity in the vehicle that could otherwise earn a return.

Of course, not everyone buys a car new. But *all* cars are bought new at the start of their life. All that happens when you buy one second-hand is that you take over the burden, allowing the previous owner to redeploy their remaining capital. Another way to understand it is to see that an older car lasts less long, requiring further capital to be found for its replacement.

Few things eat capital as fast.

In those years when I and my partner employed ourselves, we worked almost entirely at home. So we converted a spare bedroom into an office. When we filled out our tax returns, we always felt a little guilty claiming the use of a room in our home as an expense, and thus tax deductible. Until, that is, we started a family and needed another room, whereupon we found ourselves spending tens of thousands of pounds building one. This caused us to notice the value of simple space.

We realized that the biggest room we had was the garage.

18 To recap, we have already 8p/km for fuel, 1.5p on running maintenance, and 15.2p to cover servicing, depreciation, insurance, and a licence to use the public highway.

Quite apart from paying ever-rising parking charges at work and in town, as all land in a crowded area is called to earn its keep, we car owners pay big-time to park the blessed thing at home. If you lack the luxury of a garage, you must still find space outside (and face higher insurance, maintenance, and depreciation costs). More than half the homes in my street have converted their front garden into a car park. Not only do *they* lose something, everyone else in the street loses too. It shows up as a deficit in property value, especially when there is more than one car out front.

Gardens add value, visual and financial. Car parks subtract it.

A great many car owners, especially in urban areas, keep their vehicle on the street, and may think they can escape the cost of home parking. Not for long. Like it or not, the street has never been theirs to use as they will. Street parking causes significant additional risk to pedestrians and is, in any case, less than reliable, especially as the number of cars per household climbs towards two and beyond. Many local authorities are responding to the need to control street parking by introducing the infamous "residents' parking zone" (RPZ, see p. 230). Residents everywhere will soon be paying to keep their car on the public highway, in exchange for the guarantee of at least one place. That is unless the road space is needed to afford a greater flow of traffic, of one kind or another.

Parking space reflects a growing proportion of property value.

Of course, asking folk to pay for something considered free for decades was always going to cause a rumpus. At least two attempts to start charging in North Oxford have been beaten back with the usual battle-cry of "motorist persecution". But the highway *is* public property, and represents space set aside as a conduit for *everyone* to use. It is common to see a full half of the available width removed by parking. Congestion can increase as a result. Cycle-ways, which *reduce* congestion, can use that space instead.

Our additional room, of a similar size to the garage, which could have been used instead, set us back £24,000 in 2000. Again, this is money we could have invested, raising a return. It is also comparable with the difference in value of a typical family home when a garage is added.

Using the 6% rate we estimated earlier, this adds 10p to the average cost of each kilometre driven in Britain.

Let us call the sum of fuel and running wear (mainly tyres) the *immediate operating cost*, which we concluded was typically 9.5p/km (15p/mile). Next, the *cost of ownership*, including depreciation, servicing, insurance, and road-use licence (VED), which came to £2,205, or 15.2p/km (24p/mile), for an average mileage of 14,500kms (~9,000 miles). Financing both vehicle and garage adds a further 16p/km (25.5p/mile). The *full economic cost* of running a car over an average mileage thus comes to over 40p/km (65p/mile), or just under £6,000pa. This is far beyond that normally perceived.

Of course, there are ways to reduce this bill. You might run an older car and cut the cost of finance, but not by much, and you add risk whenever buying a used car, whose history will typically be largely unknown. For example, spending £5,000 three times, rather than £15,000 once, will cost only about half as much to finance, but you may find the engine worn out. You might evade regular maintenance, but only at the risk of bigger bills later, and possibly of your life. You might keep your car on the street, but only at risk of the local vandals and thieves, not to mention the weather.

In the end, the full cost will descend on you, one way or another.

Sadly, for most people, owning a car means borrowing money. As a result, a truly destructive social iniquity rises — debt.

The British population now suffers a personal debt mountain that exceeds a trillion pounds — an average per adult in excess of £24,000, or *one and a half times* their annual income. The average unsecured lending is over £4,000 per adult — a figure that grew by 10% in a single year (2005), and 45% over the previous five years. A generation has clearly grown up with no compunction with regard to borrowing, and no inclination to save.

Cars incur debt. Precisely how much is difficult to quantify. Records do not specifically correlate amount borrowed with vehicle purchase. Any such figure would be of limited use in any case because there is no way to separate personal expenditure. One might use a small inheritance to buy a car, but then, as a result, be forced to borrow money for furniture. Suffice it to say simply that running even one car is by far the biggest household expense after securing a home. Add to that the fact that 29% of households now feel the need run two vehicles — one per adult — and it becomes impossible to dissociate the car and borrowing.

Making the car a necessity forces many into a slippery decline.

Clearly, the amount we *use* our car is not restrained much by the immediate operating cost (fuel and additional wear and tear). It is even less important when someone else foots the bill. A growing proportion of vehicles on the road in Britain are "company cars", paid for by an employer, but available for private use outside hours. Not surprisingly, annual mileage rises from 20,000kms, for a car used for business or commuting, to over 33,000 — a full *two thirds* more. Currently, company cars make up only about 6% of all those on the road, but account for about 14% of their mileage.

We had already suggested that we tend to travel more to recoup our high fixed costs, given that we pay little more in doing so. When the additional cost (to the user) disappears altogether, as is often the case with a company car, so does all restraint. In fact, retailers frequently reward the *driver* for buying more fuel, even when their employer is paying. Not only is there no cost to each extra mile, there is then actually a *benefit*.

On the subject of privilege, we might briefly return to the cost of congestion. Britain is really two countries: London, and everywhere else. The average salary in London is £30,984, as I write — 40% more than the national figure of £22,411. Again, we shall err on the low side, and use the latter. Conveniently, we shall take 40% as the overhead above salary that accounts for all additional costs of employment (holiday, sick leave, office, *etc.*), recovering the London figure. If we assume forty-eight forty-hour weeks worked per year, this leaves a time value of about £16/hr. Peak time speeds in London are now just 24kph, yielding a cost of 67p/km — 68% more than the price paid to drive. It is much worse again for someone on a London salary, or rather their employer, who pays 94p/km.

A great reward awaits raising speed, or reducing our need to travel.

In town, we found that walking and cycling can move about 2,000 and 1,700pph/m[19] respectively, in each direction. The bus could achieve 2,000–3,500pph/m.

The private car represents the worst possible system with regard to capacity — uncoupled vehicles, each with low occupancy. Using our established figure for true occupancy of 1.2, accounting for the fact that a chauffeur is no passenger, we find the capacity of a single lane of 110kph motorway is

19 people per hour per metre aperture

just 1,350pph. 'Lane hogging' and the presence of larger and slower freight vehicles mean that capacity fails to rise in proportion with the number of lanes. A reasonable approximation to the truth would seem to be that a typical British six-lane inter-city motorway yields the capacity of an ideal four-lane highway — about 2,700pph in each direction. With regard to the use of land, this offers only about 225pph per metre width; less, if we were to account for the 'hard shoulder' (emergency access lane).

It would not be unreasonable to conclude that the bike can move almost as many people per hour around town using only *one tenth* the space. Alternatively, using the same space, mobility around town could increase almost tenfold.

To compare mobility *between* towns, let us wait until we assess rail.

Motorway construction is at a low ebb in Britain because the UK Highways Agency, and its mummy, the Department for Transport (DfT), have finally accepted the reality of *Downs' Law* — new roads invite new traffic. One recent project that survived is the Birmingham Northern Relief Road, now known as the M6 Toll Road. It was commissioned in 1998 and opened in December 2003, costing £485m for 43kms, *i.e.* a little over £11m/km.

Unfortunately, akin to the cost of other public projects, the cost of road construction appears to be drifting out of control in Britain. In response to parliamentary questions, it has now (2005) emerged that the cost of motorway construction has risen to an average £14.4m/km, around 27% more than that of the M6, completed less than two years previously. Curiously, the DfT do not seem to apply the same criteria in support of both road and railway projects. They are evidently more tolerant of cost inflation with road than with rail. They are also less stringent in their requirements over the balance of cost and benefit with regard to funding. Worst of all, they resist all attempts at *integration* of road and rail. The effects of the latter are plain to see. Even signs are at a premium. A strange prejudice is evident.

Taking the costs provided within the House of Commons, we obtain a cost of transport provision by motorway in excess of an appalling £5,300/pph/km. It's a similar story for local transport provision; for example, between city and satellite town or suburb. Four-lane highway now costs about £7.5m/km to construct and can carry the equivalent of only one and a half ideal lanes — around 2,000pph. This produces a cost of £3,750/pph/km.

As we shall see, while it may cost a little more to build, a railway offers far better value than road; not just in its superior capacity, but in speed, safety, and comfort as well. External costs are also far lower. A typical assessment puts the external cost of rail at around half that of a bus and less than a quarter that of the private car.[20] A car passenger also consumes up to *twenty times* the energy they would use travelling by rail.[21]

And yet, we have built only one new line here in the last century.

Whether or not you believe using a car to be *anti*social depends upon your point of view, but it can hardly be termed social. For a start, it is not shared, except occasionally, and then usually only with immediate family. In contrast, a rail station is shared by all within the local community. I have grown to know many people in my village well while using the train.

Using a train requires a degree of co-operation. With the car, on the other hand, we engage in *competition* with our neighbour. From the moment we leave home, we must compete for space on the road and then in the car park. This is why much of the sparse communication that takes place between drivers tends to be of an aggressive nature, especially in an urban environment. (I am old enough to remember the lost lamented rural "after you" protocol that has given way to the urban "after me".)

As resources diminish, human nature gives pride of place to number one. This can be true of any mode of transport, perhaps, but with the car comes a sense of anonymity. We are more willing to honk at the person in front because they are more than likely a stranger. Should we be recognized, we will usually back off. We behave differently when we are aware of being watched. Even on a crowded train, *courtesy* is expected, and we tend to behave accordingly, in a social manner.

We can hardly deny that every sin is begged by the private car. Pride, vanity, and lust, receive appeal in most advertisements, which seek to promote the idea that you can advance at least the appearance of greater importance,

---

20 A report by Union Internationale de Chemins de Fer regarding the Channel Tunnel Rail Link (2000) yielded 20, 38, and 87€/thousand pass-kms for rail, bus, and car, respectively. Although it accounted for many things, it failed to include the *largest* external cost — congestion.

21 Even when the car is capable of 40mpg, it consumes 590W/pass-km, assuming an occupancy of 1.2. A modern train can consume as little as 30W/seat-km, according to Network Rail.

allure, and potency, with each new model. Of course, the more you spend, the grander your purse, apparel, and libido, will appear.

Avarice is to be found in many a British workplace, where promotion is often seen as the means to obtain a larger or sportier company car. Given the existing technological capability to manufacture a perfectly adequate car capable of perhaps three to four times less than the average fuel consumption, gluttony is even more plainly evident. Anger and aggression is easily observed in the car behind on any occasion you feel moved to obey the speed limit, especially in the outside lane of a motorway or through rural villages. Sloth is revealed, as I write, in someone, retired but fit, who just chose to drive a hundred metres across my village to deliver a letter, rather than walk or cycle — not an uncommon choice these days.

That we are capable of sin is something of which we are all aware but prefer not to contemplate. Over the ages, much of social structure has been deliberately fabricated to help us resist the short slide back to the cave. Rome was as well aware of what would happen without faith and religion as it was of the power it could wield with them. But labour, disease, and poverty, worked equally well at keeping the lid on. Now it is different. In the rich world, the grip of religion has failed, and leisure, health, and prosperity, abound. The lid is off, except perhaps for personal mores and conviction, which take time to grow, where they can. Meanwhile, we go among each other more often, with less consideration, and with fewer principles.

It would be absurd to blame the car for all human sin, but it is undoubtedly part result, part symptom, and part cause, of widespread social decay.

Whatever its advantages, the car is *not* ultimately a necessity. While nearly a third of all British households run two or more cars, a quarter still run none. This rises to more than a third in London, which is better served with alternatives than anywhere else in the country. It is thus relatively easy to avoid car dependence there. Transport is regulated and planned effectively, in a way that is democratically accountable, in principle at least.

The London Congestion Charge has seen congestion reduce by about a third, and cycling rise by a third. Bikes are returning as a common sight after decades of absence. If this is properly supported by the construction of safe cycle-ways and cycle parking, car-free living will become easier still.

London exhibits an array of relatively self-contained foci, interconnected by a comprehensive subterranean light-rail network. Those two observations are critically connected. Transport and land use are two sides of the same coin. You cannot decide one without the other. London was spared wholesale demolition and strangulation by road congestion in the 1960s by a few who understood enough in time. They arrived at their conclusions scientifically. Yet there was another — an American genius — who, at about the same time, understood just as much by sheer instinct.

But that's a story we shall keep for our final chapter.

Rail

In a world today almost bored by innovation, it is hard to imagine the impact the railway had on its arrival. The locomotive was the one machine that moved outdoors. At a time when few ever ventured into town or even to the neighbouring village, it could take you to see the sea. The sheer power of a steam engine — something it did nothing to hide — left no witness unaffected. Control of such power invoked dreams in every boy, large or small. Only the rocket would eventually surpass this. But rockets don't fly past your window at regular intervals, nor do their drivers sometimes let you sneak aboard.

Steam engines were living breathing dragons comprising every element — earth, air, fire, and water — but trains came to be symbols of so much more. Railways, not roads, opened up continents, enabling one in particular to become a single nation. America may now be made of cars, but it was made *by* railways. Only trains can deliver troops in sufficient numbers, with sufficient precision, to remake warfare. They delivered Hitler's divisions from west to east, wherefrom few returned. Slaves came to hear freedom call in the three-tone whistle of the trains they would one day ride to Chicago, taking their Blues along with them. On the way, they collected the boogie rhythm that would one day grow into rock n' roll.

Trains brought death, liberty, and music.

Time came too. No one had needed a clock in Bristol to read the same as in London, until Isambard Kingdom Brunel joined the two by rail. Stephenson, Brunel, and others marched their iron and navvies across

Britain, chalking up around thirty-seven thousand kilometres by the start of the First World War. Most would now be surprised to learn how close the railway came to them. Every settlement of over 20,000 had a station, 98% of those over 5,000, and nearly two thirds with just 500 souls.

Almost the entire population of Britain lived within a bike ride.

Those who started what remains the greatest transport revolution in history were no fools. They saw three tricks that would dramatically increase speed and efficiency, and thus their fortunes. First was to use the new technology of iron making to avoid the hideous difficulty and expense of laying a road. Instead of digging up the ground, you lay iron rails over the top of it. This was to prove much harder and more expensive than it sounded. Brunel would drive piles to support the rails. They were so far apart that the rails would sag in between, giving an uncomfortable 'switchback' ride. Soon, the problem would be solved, and a degree of comfort bestowed unmatched before or since. But an even greater benefit would follow.

An iron wheel on an iron rail has about two thirds less rolling resistance than a rubber tyre on a road. Less than a third the energy is needed to move the same mass.

On the road, increasing speed does *not* raise the number of people or packages you can carry per hour, simply because the gap between successive vehicles must increase. In fact, because that gap increases out of proportion with speed, flow actually decreases (see Figures 2.6–7). This is not what happens in a river. Perhaps that inspired someone. Or perhaps it was just that steam engines were incredibly expensive and it was obviously imperative they should each haul as much as possible. In which case, the idea of equipping each wagon with its own engine and driver, and having them follow each other a safe distance apart would appear just plain daft. History might have unfolded differently had not steam been so very powerful.

Whatever the origin of the idea, coupling carriages together to form a train breaks the link whereby greater speed means less flow (Figure 2.8). In fact, it can permit speed to *increase* flow. Fewer services can carry as many people, when more are served each time. Lower departure frequency is exchanged for higher occupancy.

A larger separation between vehicles (trains) allows them to safely travel at greater speed. Greater speed will attract more passengers, to fill even longer trains, at a higher price. The particular strength of the train has always been in inter-city travel, where a plentiful supply of wealthy passengers can be found at either end.

To early rail entrepreneurs, it must have seemed too good to be true. People, who had never travelled faster or further than a horse could propel them, took to the trains in droves, often out of mere curiosity. Proprietors quickly realized how greater speed could attract more custom, and the race for pace began, culminating in Mallard's wondrous 126mph (201kph).[22]

There is one more benefit rail brings, of particular relevance today.

When the railway began, there was much track and few locomotives. Each train could be assured of an uninterrupted path along the entire "permanent way". As popularity and traffic grew, with slow freight mixing with fast passenger trains, it may have seemed obvious what was required. Instead of relying upon *ad hoc* signalling and simply putting up with frequent delays, faster passage and a guaranteed arrival time could be had by using a strict *timetable*. In other words, work out a practicable schedule in advance, and run the network like clockwork.

Allocation of a continuous unimpeded path to every service is the third great trick of the railway. It affords the twin gains of speed and predictability, with greatly increased capacity and efficiency thrown in for free.

Today, with serious and worsening road congestion, the ability of the railway to offer short and predictable journey times into town is invaluable. London could never enjoy such prosperity without it. While the sea founded its fortune, the railway multiplied it. To look upon transport as simply another independent business is to miss the point entirely.

Any beating heart needs adequate arteries.

A railway can offer arguably the highest quality of any transport mode. A train can be very fast, provided there are not too many connections, and very comfortable, provided it is not overcrowded. You can work, rest, or play *en route*. The most common objection is the high construction cost.

Building railways has never been for the financially faint-hearted. When Brunel built the Great Western Railway (GWR) his annual salary of £2,000 allowed him to buy a Duke's mansion in London and employ a private reti-

22 A record that to this day remains unbroken by any steam train.

nue. His personal assistant even had two assistants of his own. Merely the first phase of the GWR, from London to Bristol, cost nearly £6,500,000.

Aside from the engineering, Brunel had to raise finance in the face of enormous risk. But the payback to the nation was incalculable. The wealth of the richest empire in history depended on such railways.

There was no serious competition. Roads were passable only for light loads. Freight depended on canals, which could boast respectable capacity, but were terribly slow. Railways, however, were far cheaper to build, much faster, and could carry greater loads; they were the 'broadband' of transport. They could also penetrate the country in ways impossible for canal builders, eventually reaching every last town and village. An economic boom followed in their wake, along with mobility beyond the dreams of mortal men.

Today, every railway faces serious competition, for both freight and passengers. Construction cost and risk are even greater now. Like Brunel, a modern railway builder must to exploit the latest technology. The "state of the art" is the French *Ligne à Grande Vitesse* (LGV). Only one such line has thus far been built in Britain. Section One of the 109km Channel Tunnel Rail Link (CTRL) opened in September 2003. Section Two finally connected Britain to the European LGV network four years later.

The price ran to around £5,200m, or £48m/km. Even disregarding the premium charged for laying a quarter of the line underground, no railway has ever cost anywhere near as much per kilometre. Given that tunnelling is roughly five times more expensive, surface construction would have come in at around £23m/km — half as much again as the LGV built recently from Paris to the Mediterranean. A Spanish LGV opened the same day as CTRL Section One, but at a price *three times* lower per surface-kilometre.

Many explanations are possible for the high cost of LGV delivery in Britain. One is excessive bureaucracy and regulation (embarrassing, if true, since it is a common accusation levied by the British against their European neighbours). More likely is that building just *one* line, rather than an entire new network, carries a high overhead. Other countries are able to save costs dramatically through greater overall investment in equipment, production facilities, and expertise. This might also explain why the mere upgrade of the British West Coast Main Line (WCML) — raising speed by just 40kph — eventually cost £16m/km. It might even have been cheaper to pay the French to build a brand new LGV.

A coherent plan for the long term would appear critical.

As we have noted, greater speed does not confer greater flow but likely the reverse. As speed rises, the frequency of departure must fall. The only way 300kph *Trains à Grande Vitesse* (TGVs) can retain capacity is by dramatically raising their occupancy. Each Eurostar is typically *twice* the length of a standard inter-city train. With eighteen carriages, it can carry *nine hundred* passengers. Its 400m length presents no serious operating problem; it just needs well designed stations that guide you to your seat.

As well as greater speed, greater safety has added value to a modern rail journey. In twenty-seven years of network operation, and over a hundred billion passenger-kilometres, not one TGV passenger has been either killed or seriously injured. Then there is the comfort and equity of access of rail.

A twin-track railway can sustain the movement of at least 150,000 people per hour (pph) in each direction, leaving a capacity cost of around £115/pph/km. The figure reflects, as before, only half the possible occupancy.

This makes rail more than *forty times cheaper* than road in its capital cost of transport provision, while offering three times the speed.[23] Safety is at least twenty times better on rail than road. Time spent driving is simply lost, whereas it can be put to good use on the train, by working *en route*, for example. Many other differences could also be considered, like the difficulty, and delay, in gaining access to refreshment or a toilet.

Much was lost when we exchanged rail for the privacy of a car.

Not every region is equally suited to the full might of LGV. France is well suited with many large conurbations separated by hundreds of kilometres of sparsely populated countryside. Britain is very different. Cities are often clustered, rather than spread out over an area. The bulk of the population lives within 200kms of London — a radius within which LGV is typically less appropriate.[24] London is also surrounded by deep and densely populated

23 The successor to TGV, the AGV, or *Automotrice à Grande Vitesse*, is scheduled to enter service in the summer of 2011, with a speed of 360kph, while consuming 30% less energy.

24 Even when ignoring the need to stop frequently at intermediate destinations, the reduction in journey time may not warrant the increase in expense. A greater frequency of service is also commonly expected, the shorter the distance.

suburbs, which is why so much of the CTRL was built underground, raising the cost dramatically. A similar escalation results where, as is often the case in America and Australia, inter-city distances run beyond a thousand kilometres. But even in those countries, there are clusters of well-separated dense conurbations that could be well, and economically, served.

As with the bicycle, though on a much larger scale, *focal* development is the key. Sprawl is the enemy.

Rail is better suited to running underground than road. Extended tunnels demonstrate much higher safety. As a result, it can be unreasonable to count the cost of tunnelling against rail. Once a tunnel is essential, or just preferred, rail is usually far the best option.

Where tunnelling cost or surface noise prohibits, existing conduits may be restricted to conventional speeds of 200kph with more frequent, shorter, trains. Even then capacity can exceed 100,000pph, with a unit cost of about £160/pph/km. We might also consider two other categories of railway: district 'transit', with a speed of around 100kph over distances of ten or twenty kilometres per stop, and local light rail at, say, 50kph, with stops only a kilometre or two apart. These can each yield capacities of 70–80,000pph, equating to about 10,000pph per metre aperture in each direction (again, assuming just half the seats taken).

Capital costs of light rail vary dramatically according to environment, but in exactly the same fashion as those of road. Land use, together with the cost of increased tunnelling, bridging, and moving drains, water pipes *etc.*, make urban conduits far more expensive. Here, we shall take more interest in the cheaper systems that serve suburban and district routes. These currently come in at around £4.5m/km, giving a unit capital cost of about £70/pph/km — *fifty times* less than a comparable road, used predominantly by the private car, and half the price of a dedicated bus path.

Rail clearly represents far the best value in the provision of new transport infrastructure at any scale short of the intercontinental.

Railways also last a very long time. Although ballast, sleepers, and rails have been replaced on the GWR many times, the track-bed is essentially the one Brunel's navvies laid one hundred and seventy years ago. The Channel Tunnel, which could only be safely and economically exploited by rail, will undoubtedly remain useful for centuries. The financial problems endured by EuroTunnel are at least in part due to the failure of ordinary financial instruments to reflect that fact. Shareholders need worthwhile returns before the guy with the scythe comes knocking.

Rail forms the classic example of a marginal cost that *falls* with demand, in stark contrast with its competition, the automobile on the public road. To haul a single carriage, there is all the expense of establishing conduit, track, and signalling. From then on, things get better quickly. As demand rises, it costs little extra to run more frequent services, reducing delay in departure and in making a connection. Trains can then be made longer, and even a second deck introduced, increasing occupancy. In continental Europe, twin-deck trains are in widespread use (Figure 2.9), for suburban, district, and regional services.

Figure 2.9 A twin-deck train in the Netherlands.

It is simply far easier to raise the occupancy of a train than of any uncoupled vehicle. Doing so is relatively inexpensive. Raising occupancy increases capacity and reduces ticket price.

A train's length can vary according to the kind of service required. Where journeys are typically short, the wait at the station can matter most. More frequent, but shorter, trains can carry as many people per hour. (Stations can then be shorter and thus cheaper.) The shorter interval commands a lower speed, but this is limited anyway, on a local service, by more frequent stops. For longer journeys, speed is more important than delay, so longer trains provide less frequent but faster service.

These things fall together with trains. They fall apart without them.

There is a common perception of rail as the "technology of yesteryear". While this may be understandable, it is utterly wrong. For example, there is far greater scope for the *automation* of a railway than there is for road transport.

Keeping lines running to capacity requires 'moving-block' signalling, where a safe interval (block) is maintained between trains as they move along. This is hard to implement with the degree of reliability and fault tolerance necessary, yet it has the potential to *double* the value of any railway — an extremely expensive asset, with a lifetime measured in centuries. Moving-block signalling is already in use on light-rail systems, such as the London Docklands Light Railway (DLR) and Paris Metro. For main lines, the EU is introducing a standard that incorporates it, as part of the European Train Control System (ETCS).

Information technology can render rail investment intrinsically even more attractive. Trains on the aforementioned DLR, for example, need no driver. ETCS will make the train driver redundant throughout Europe.

Little boys will have to dream of some other occupation.

Sadly, in Britain, massive inefficiency remains in the maintenance of railway infrastructure. A 2006 TV documentary portrayed archaic restrictive practices, gross mismanagement, and incompetence, on an horrific scale. It concluded that, of every eight hours work scheduled, only around one is done.

In 2002, estimates for electrification ran to £4.8m per kilometre In 1986, allowing for inflation, the cost was just £1.1m. Similar cost escalation can be found in the laying of track (reinstating twin-track 'singled' in the 1960s) and in line upgrading (*e.g.* the aforementioned WCML). At the same time, some privately run 'heritage' railways, that have no option *but* to control costs, achieve profitability without subsidy, despite their dependence upon a short tourist season each year.

In the hands of the State, track investment has been sporadic, and at best piecemeal. This pattern fosters the popular British misconception that railways belong in the past, despite overwhelming evidence to the contrary abroad. Other consequences of state provision include the failure to relate charges to costs incurred, and the almost total absence of research and development, despite its potential payoff.

There remains the question as to whether railways truly *need* state subsidy. Could they once again become self-sustaining businesses?

In 2004, track access charges were levied on train operators in Britain at the rate of about £2.75 per train-kilometre, which fell on passengers at the rate of about 3p per passenger-kilometre. This raised £1,287m, whereas the State spent £5,105m on track "operation, maintenance, and renewal" (OMR); the difference being subsidy raised from general taxation. (Omr expenditure was dramatically inflated at that time because long-term neglect had led to an enormous backlog of renewal.[25])

Only £3m (0.06%) was spent on research and development.

Given the evident inability of the State to control costs effectively, it does not seem unreasonable to suggest that an efficient private operator could maintain the infrastructure with the income from current access charges alone. Recall that cost escalation over sixteen years amounted to in excess of a factor four, accounting for inflation. The contribution to the eventual ticket price need be no more than 3p/km.[26]

In the real world, every capital asset must yield a return. This should surely include the railway network, whose construction required tremendous labour, land, and enterprise. A return of 5% on capital valued at, say, £5m per track-kilometre would require another 13p/km from each passenger. But this would assume the construction of a brand new network, which would then be worth a total of £165bn, if it were the same size as the existing one. Either we should be content with a *much* lower return, because it continues for very much longer than most alternative investments — making 5% overly generous — or we should considerably reduce the valuation, in the same respect. Bean counters usually favour the latter approach. Our 19,000 route-kms are currently valued at just £23bn. A return per passenger (including depreciation) might thus be less than 2p/km.

Should road-use charging arrive, a similar argument would apply.

Only operating costs remain, consisting of train depreciation and maintenance, stations, staff, energy, and overheads. By far the largest of these is depreciation. A new Eurostar set costs around £10m but it takes a lot of looking after. Leasing is now standard practice in Britain, and gives a way of at least putting an upper bound on the cost of "rolling stock". A two-carriage diesel-multiple-unit (dmu) can be had for £260k *per annum*

25 The extent of dilapidation was brought to public attention by a tragic calamity in October 2000. A cracked rail caused an express to crash near Hatfield, killing four. The entire British rail network was reduced to low-speed operation for several months while track was checked and renewed all around the country.

26 We assume charging about the same to carry a passenger as a tonne of freight. Similar numbers of each typically fit on a railcar. Arguably, freight should pay more, as it incurs greater wear on the track and must travel more slowly in respect of safety.

(pa), though the leasing companies set up after privatization are far from competitive.[27] On local lines, such a train might clock up only around 100,000kms each year, which, with half-occupancy, comes to around 6.5p/pass-km. Larger, faster, trains log much higher mileage with many more passengers, bringing this cost down to around 2p.

An average of around 4p/pass-km thus seems a reasonable contribution towards vehicle finance. We shall allow another 3p for remaining costs (driver, fuel, cleaning, *etc.*).

In summary, we proposed 5p/pass-km in return for the track, and 7p for the train. There remains the need to consider the provision and maintenance of stations, operating overheads (information services, ticketing *etc.*). For this, we might allow a further 2p, bringing the total cost of delivering a rail passenger-kilometre to around 14p.

Obviously, there will be considerable variation with the quality of the line and the rate at which passengers flow over it. We have considered only an average. A higher-quality line, with faster operation, will see a lower flow and higher costs. But then a low-speed rural line, while it may cost less to build and maintain, may suffer low demand. A charge based on a network average does not seem unreasonable.

Current (2007) UK direct ticket price varies from 11p/km for a local return journey (30kms, peak-time), through 12p/km inter-city within district (100kms, off-peak) down to 9p/km within region (630kms).

While current charges clearly do not cover costs, the difference is not extreme. By reintroducing proper cost control and investment, and by making best use of new technology, it is arguable that private provision of track could indeed become fully economic.

Taxes could then be saved, market accountability restored, and politics removed. Research and development might even recommence.

Enterprise would be free to return.

After sterling service during both wars, the British railway became what it remains today — "nobody's baby". Much loved by many, but by none with resources sufficient for investment, it fell into decay. State provision, as

27 In January 2004, the Chair of the UK House of Commons Select Committee for Transport, Gwynneth Dunwoody, urged an investigation, saying that leasing companies were costing the taxpayer "bags of gold".

opposed to mere regulation, then turned it into a political football, with a reorganization, and its consequent disillusion and disaffection, after almost every change of government.

Meanwhile, competition with road transport continued to grow. A critical, and fundamental imbalance was in liability to track cost. While every train had to pay in full for each kilometre run, cars and trucks paid only a fixed fee for use of the road. Charging per road-kilometre, according to demand, has often been proposed but is always kicked safely into the distant future, to be introduced piecemeal if at all, as the inevitable political backlash allows. Much less public, was the *two and a half fold* increase in track charges upon rail operators in the five years to 2009. Yet they still possess no control whatsoever over the inflated costs these enclose.

Price increases were inevitable and steep. Many services may now prove uneconomic, depositing more traffic on roads already congested.

Relating price to marginal cost[28] is critical and is what the private sector inherently does well. The need to exploit a market to the full this way is precisely why early operators began offering Second and Third Class seats. (Brunel originally envisioned only the gentry riding his railway, simply because he expected the full economic cost per seat to fall beyond the means of anyone else.) But there is to be no flexibility in track access charging. The small passenger train, that imposes a tiny additional track cost, simply won't run, or will run so infrequently that precious few will use it.

Like any other business, railways need enterprise.

One cogent argument for the State to support railways is that transport is unlike other industry, in that it provides *infrastructure*, both for other industry and for our lifestyle. After all, the railway *began* the idea of travel for ordinary people.

Curiously, the idea of transport as infrastructure has been rejected as resolutely by a socialist government as it was by Margaret Thatcher. Despite this position, both invested heavily in road, but not in rail.

Other nations are fully committed to the idea of transport as infrastructure, and invest heavily on purely economic grounds. Given the success of TGV and its cousins, Europe is developing an integrated LGV network. Services

28 The marginal cost here is that of carrying each *additional* passenger.

between London and Amsterdam, and into Germany, are expected within a few years. Spain has tabled a €250bn investment to bring all its citizens within 50kms of a TGV station. Such mobility, together with the ability to work *en route*, will confer real economic advantage. Meanwhile, Britain has yet to entertain the building of any LGV beyond the CTRL. It is in grave danger of being stuck at the end of its very own branch line.

We had better hope that Europe will prefer to keep such things open.

Thus far, we have made little mention of the environment, a full discussion of which is left for Chapter 4, but it remains worth making three points here. First, modern trains, like TGV, are electrically propelled, removing them from the debate on energy sources and carbon emission. Second, while they still consume energy, they consume far less per passenger-kilometre than any automobile. Third, the LGV network has thwarted growth in 'short-haul' aviation within France, along with its significant carbon emission.

Almost all the growth in aviation is short-haul, mostly over land, in direct competition with rail. The combination of TGV speed with the traditional ability of rail to deliver us in the centre of town, and thus avoid the infamous airport delay and hassle, has proved decisive.

In Britain, it is a different story. In January 2007, The Independent reported that a return rail ticket from London to Manchester cost £219. The competing flight cost just £171. For the price of the rail ticket, one could fly to New York and back — twenty times as far. This is despite the fact that many more seats are available on the train to Manchester (~17,000/day) than on the plane (~10,000/day). Something has clearly gone wrong.

Rail costs need proper control, and *all* aviation costs must be met.

One justification for the State to intervene and stimulate a railway renaissance, as it has in France, is that flying liberates around *five times* the carbon per passenger-kilometre. The difference is profound.

Aside from a responsibility for national infrastructure, the British Government now has (legally enforced) carbon emission targets to meet.

Neither of these motivations should form an excuse for the State to continue in railway *provision*. After all, if the State is already providing, and failing to be competitive, how can this be part of any solution to any problem?

Regulation alone can achieve a great deal.

In the late 19th century the British Government limited railway operators to a fixed rate of return on their investment. This meant that companies could not increase profits by the usual method of increasing efficiency or providing a better service. The only way was to expand the asset base. So they built hundreds of barely economic lines out into the distant corners of the British Isles, bringing service to everyone.

Today, there are many ways the State could aid the development of the railway without being involved in provision. For example, stations could be treated as the civic amenities they are, and receive public investment to make them more accessible and attractive. Aside from this, and allowing operators to control *all* their costs, surely the greatest assistance would be to force competing modes to meet the same standard of safety.

On the 23rd February 2007, a Virgin Pendolino express train derailed at 95mph. While the cause of the accident was culpable negligence, and unforgivable as such, the consequences were far less severe than they would have been either in aviation or on the road. It is still terribly sad that one person died and twenty-two were injured. But if rail carriages were not designed to withstand such events, many more would have lost life and limb. Building them to that standard is very expensive.

Between that rail fatality and its predecessor, more than *ten thousand* died on the road. Failing to make the car as safe is even more expensive.

Summary

We have considered all the alternatives available to us for overland transport over a distance up to a thousand kilometres — walking, cycling, bus, train, and car — remembering that the vast majority of journeys are *local*.

The reader could be forgiven for perhaps feeling a little overwhelmed, and not a little confused, by now. It is surely worth attempting to compare these options in one place, according to at least some of the criteria we previously established — speed, price, safety, capacity, and the cost of provision.

Table 2.1 offers a quantitative comparison between modes.

Some other factors are clearly as important but less easy to quantify, such as privacy, convenience, and the effect upon landscape and society.

parameter	unit	foot	bike	car	bus		rail		
					local	*coach*	*light*	*heavy*	*lgv*
range [1]	kms	2	10	30/700	20	700	40	1,200	1,800
speed [2]	kph	3	15	50/110	30	110	60	200	300
casualty rate [3] *(no car)*	deaths/bn pass-km	0?	1?	n/a	?		0 [4]		0 [5]
casualty rate (with car)		52	39	5 [6]	0.4 [4]		0.3 [7]		0
price [8]	pence/ pass-km	0	<1	41	42-65	11-22	14	14	14
capacity [9]	pph	2,700	2,400	1,350	26,000 [10]	27,000	60,000	100,000	150,000
capacity per unit aperture	pph/m	1,800	1,600	225	3,500 [10]	3,500	7,500	14,000	20,000
cost of provision [11]	£/pph/km	100	115	2-5,000	160 [10]	250	70	160	115

TABLE 2.1 A quantitative comparison between transport modes.

1 Estimated for 40mins travel on foot, bike, local bus, or light rail; 6hrs travel otherwise.

2 Track speed only; no account taken of departure delay, stopping *en route*, bunching, or making a connection. Does *not* reflect actual journey time.

3 Inferred by information published by UK DfT for Britain in 2004.

4 No distinction in records between local and national services.

5 No deaths yet on either UK CTRL or on French LGV network. Ignores Inter-City Express accident in Germany caused by design flaw absent in both Eurostar and TGV fleets.

6 Assumes vehicle occupancy of just 1.2, and equal risk in freight mileage. Includes deaths both suffered and incurred by all mechanized road transport.

7 Single accident caused by car on level crossing at Ufton Nervet, Berkshire, causing seven deaths including driver of both car and train. Accounts only for rail passengers killed.

8 Track and external costs are made highly uncertain because of state provision (of both road and railway). See text for discussion. Otherwise, refers to full economic cost.

9 Estimated maximum duplex flow for typical conduit — two-lane cycle-way, twin-track railway, six-lane highway.

10 Accounts for either bunching or ticket collection *en route*, but not both, in reducing effective speed to just 12kph. Accounting for both would reduce this to 6kph. Capacity would then be little more than half that shown. Cost of provision would rise accordingly.

11 Road construction costs (2005) were provided in answer to questions in the UK House of Commons. The figure quoted refers to four-lane highway. Cycle-way cost was indicated by recent (2008) construction in Devon. Pavement (footway) cost is assumed identical. Rail estimates are for surface construction of line only, and assume competitive tender.

Recall that around 80% of our journeys are of less than ten kilometres, and about 90% are of less than twenty, and you can see how important walking and cycling are. They are cheap to resource, almost free to use, and, in the absence of the car, almost completely safe. In addition, they make good use of the little space they occupy and cause no pollution.

To provide instead for the car will cost at least twenty times as much, and consume seven times the space. With regard to value, journey time will rarely be less than with a bike because of congestion (see next section), while the price of travel will rise to a level that can only be called exotic.

Walking and cycling then become wholly unsafe, with eighteen times as many dying outside a car as inside one, per kilometre travelled.

Beyond a 10km radius, rail can be seen to dominate as a contender for investment, offering twice the speed for half the cost of the bus, over district/suburban distances (within about an hour's travel). It also offers twice the flow through the same aperture, where land remains expensive.

Rail retains its dominant position with inter-city travel. The LGV brings an entire nation within three hours travel. We can now span a continent within half a day, and remain free to work, or rest, the whole time.

When we consider energy and pollution, the case for rail over road will become even stronger still.

2.3 Competition

Speed and flow

Any transport service, whatever the mode, can be regarded as a product, just like anything else. As with any other product, demand can outstrip supply. When that happens, a queue forms. In the case of transport, the path concerned suffers 'saturation' — more commonly known as *congestion.*

Since value declines with the degree of congestion (the journey takes longer), while the price rises (you burn more fuel), you might wonder why it ever occurs. After all, we learn at school that if supply cannot rise to meet demand, then demand must fall to meet supply. No one complains about a 'shortage' of diamonds. Their price reflects their scarcity. No queue forms outside the jeweller, on that account. Queuing for a transport service can only mean that its price is below what the market would command.

Clearly, the market has been prevented from operating.

A deeper mystery is why, when a congested conduit is widened, speed often remains the same, or even falls. An explanation has been around for a long time but is only now becoming fully accepted.

First, we need a little background.

Typically, the availability of space for transport is limited. For example, in a mediæval European city, streets are narrow. The aperture through which traffic, of any kind, can move is narrow. The problem is exacerbated by modes that require extra space for parking and/or manoeuvring. Any such area, together with the conduit itself, is called the 'land-take'.

The *flow* of traffic passing through an aperture is best measured by counting people (or tonnes of freight), rather than vehicles. Measuring the flow through a transport conduit can be awkward, as there may be frequent junctions and tributaries. We must count the people who pass through both entrance and exit.

The *capacity* of a conduit is the greatest flow it can sustain.

There are three factors that decide flow, and thus capacity: *occupancy* (passengers per vehicle), *frequency* (vehicles per kilometre of traffic), and *speed* (kilometres of traffic passing per hour). Multiply these three together and you count passengers per hour, *i.e.* the flow :

flow	=	*occupancy*	×	*frequency*	×	*speed*
people/hr		passengers/vehicle		vehicles/km		kms/hr

When the result is a maximum, you have the conduit capacity.

Should we want to raise the capacity of some path then we have to address our attention to one or other factor, independently at first.

Raising speed *seems* like a clever way to raise capacity, since it also raises value, and thus ticket price, but three difficulties arise. First, it is likely to prove expensive; even a 10% rise in speed can sometimes increase cost by 50% or more. Second, it may well call safety into question. Regulatory authorities, and perhaps customers, will not be easily convinced. Third, and worst of all, if the conduit is already full then increasing speed forces a *disproportionate* reduction in vehicle frequency. The separation between vehicles must grow rapidly with their speed, as we saw earlier.

Capacity would actually decrease.

Conversely, any increase in vehicle frequency will force a corresponding reduction in speed. Speed and frequency are intimately linked.

One trick does remain open, however, by which speed may increase without reducing capacity. If passengers are willing to accept a lower frequency of *service* — fewer departures each day — then a smaller number of larger vehicles can be used to accommodate the same traffic. In other words, we can exchange frequency for occupancy, leaving capacity undiminished. Since conduits usually have limited aperture, this will typically mean longer, rather than wider, vehicles. One way, of course, to achieve a really long vehicle is to couple carriages together to form a train. The cost of raising speed can then be offset by selling more seats at a higher price.

Trading frequency of departure for speed becomes less attractive as the trip grows shorter, and thus more frequently made. Waiting around for a departure rapidly becomes unacceptable.

But then we only really need speed for longer trips.

This helps to explain the particular appeal of both train and car.

Value in a transport conduit

Transport conduits cost a lot of money. Therefore, they must represent a sound investment, whether that is made through private enterprise or by public agency. To calculate the return, an accountant would add the value to each client (traveller) joining the conduit to a cumulative sum. That value may rise or fall as others join.

Value, like beauty, lies in the eye of the beholder. For the traveller, it seems that value is principally found in their journey time. The shorter the journey time, the higher the value placed upon that particular route and mode. There is considerable empirical evidence that ticket price must rise dramatically to offset any saving in time. This makes perfect sense where pay is high and thus time precious.

It is more common to speak of the *cost* of using a path, rather than its value. (The one is just the inverse of the other.) This harks back to an economic analysis made a long time ago, with regard to the carriage of freight, where time clearly reflected operating cost.

The very first traveller to arrive on a given path will find some limit to their speed, even though they have the thing to themselves. Perhaps it will be set by the relevant authority with safety in mind, or perhaps it will be decided simply by the limitations of their vehicle. As more people join, subject only to the same speed limit, revenue will obviously grow. More tickets can be sold, at the same price. It will continue to grow until the conduit is full, or 'saturated', by which we mean that speed is limited by the number of vehicles upon it, rather than by any legal or physical limitation — *i.e.* the guy in front is the problem, not the speed cops, or the fact that you can't afford a Ferrari. What happens from now on is less simple.

As more customers continue to join the path, after it has become full, speed will begin to diminish. The cost of using it (your time) thus rises. New customers continue to arrive and buy tickets, but the price they are willing to pay will fall. Worse, each additional vehicle diminishes the speed of *everyone* already there. Next time they arrive at the toll-gate, they too will offer less. As the number of vehicles per kilometre (frequency) rises, there will come a point at which the total revenue begins to decline.

In extremis, speed can fall to zero in theory, and very nearly so in practice, whereupon the conduit has no value at all, to anyone. A point soon after speed begins to fall is ideal, in that it maximizes revenue, and thus the value of the asset.

Revenue is just the flow (people per hour) multiplied by speed (the main influence on ticket price) :

$$\textit{value} \;=\; \textit{flow} \;\times\; \textit{speed}$$

As more vehicles arrive, flow peaks *later* than value. Both speed and value start to fall in direct proportion with their number, while flow remains roughly the same. We must decide whether we desire the best value from an asset or merely the greatest flow. Anyone whose priority is the latter will tolerate congestion. They will represent neither customer nor operator, who want speed and return respectively.

Every transport conduit has a characteristic vehicle frequency that maximizes the return on investment. It is therefore essential to control the rate at which vehicles join a path. Price must vary accordingly.

A failure to charge for access will destroy value in any conduit.

Why you cannot build your way out of congestion[29]

One Anthony Downs, an American economist, put the cat among the pigeons, way back in 1962, by suggesting that further investment in a saturated conduit will always fail to relieve its congestion. In so doing, he offered the beginnings of an explanation for the observation we remarked upon earlier — that the speed of traffic in towns stays the same regardless of road investment.

Downs' Law	Any additional capacity provided will be met with induced demand.

In Downs' own words :

> "On urban commuter expressways, peak-hour traffic congestion rises to meet maximum capacity. ... route shifting onto the expressway will

29 This section may prove a little tough for those with no background in a mathematical discipline. The author hopes that the gist will get across, even if a nuance goes stray.

> continue until the average speed is reduced to below the average speed on alternative routes."

This he predicted because the 'expressway' offered a shorter path. Therefore, even at the same speed, the user still has an incentive to "shift route". An equilibrium will be reached between the 'slow' road and the 'fast' only when making the complete journey using the latter takes as long. Downs referred to this as "triple convergence". At peak hours, traffic would be *induced* onto a new 'expressway' from three sources — other routes, other times, and other modes — until no advantage remains apparent.

Underlying all this is a question of economics: what happens when two services compete and the cost perceived by the user of one rises more steeply than with another? This has been a focus of interest among economists since the 1920s.

With regard to transport, we need not care what modes are employed. We can forget *how* we travel for now, and just worry about unit cost and the flow on a particular route. Remember, cost reflects principally journey time.

Think of a fast but *narrow* road — perhaps a "back route" that offers a short cut — in competition with a slow but *broad* one. Assume the ticket price to be the same in both cases, and suppose the number of travellers on this route starts small and then grows.

In Figure 2.10, the black dot represents the cost (journey time) shared by *all* travelling along one or other conduit. As demand (flow) rises, the 'trajectory' of cost moves from that of the narrow path onto that of the broad.

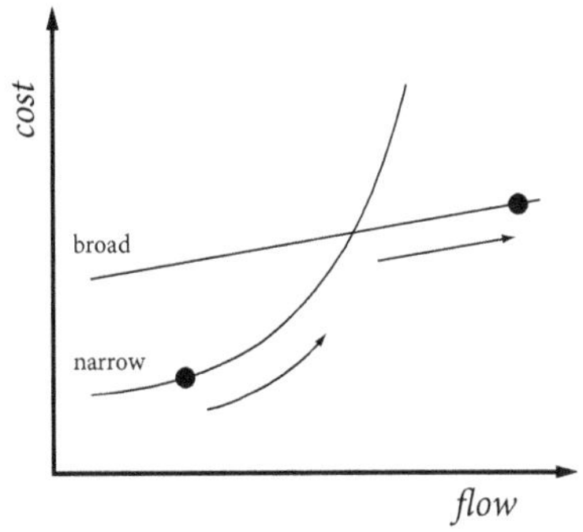

FIGURE 2.10 Cost trajectory when broad and narrow paths compete.

We would expect travellers to opt for the 'fast' path at the outset, when they are few in number. As their number grows, there comes a point where *all* would benefit were additional travellers to use the 'slow' path instead. If they continued joining the 'fast' one, it would become slower than the broad, worsening the journey time of everyone already on it.

Downs merely turned the argument around to account for the common phenomena of the time, which was the building of short cut roads with the intention of decreasing journey time. The result was, as he described, migration onto the new conduit, induced by the prospect of time saved. This would continue until the cost of using each conduit was the same

Here was the explanation for the "constant speed" phenomenon.

Imagine Figure 2.10 with only the wide, slow, path, with the black dot in the same place, representing the same cost and flow. Then the new narrow, but fast, path is added. Many migrate onto it, until its advantage disappears. The black dot will end up almost where it was before.

The additional capacity provided by the new road might allow some slight relief, but for additional demand induced from other paths and other times, as Downs predicted. Then we end up *exactly* where we were before.

What Downs failed to predict is that some new demand induced might be due to *additional* travel, and not just migration. The new expressway then not only fails to "relieve congestion", it actually exacerbates the problem by encouraging an increase in traffic overall.

It is hardly surprising that, when any new highway clogs up, travellers campaign for it to be widened. To those who are unaware of Downs' Law, yet more space appears the obvious solution. Odd though it may seem, the best investment is to speed up the slower path, as traffic will shift to it earlier from a faster one. *Both* then gain in value. The narrow path gets faster, and the broad one sees more trade.

It has taken a very long time for Downs' Law to gain acceptance. Though he proposed it back in 1962, only in the late 1990s did the official advisory body to the UK Government Department for Transport admit that it "defies credibility to deny the existence of induced traffic".

Pricing by marginal cost

When someone joins a path that is already full, they will perceive only the consequent *average* cost, measured, at least in part, by the time it takes to traverse it. However, in joining, they have incurred an additional cost to everyone already there. All traffic will slow a little. The change perceived by each individual traveller may be small, but the effect is cumulative.

The total marginal cost incurred may be considerable.

Were the operator of the conduit to levy a charge on each new traveller equal to the marginal cost of carriage, traffic would migrate earlier to a slower alternative. More space would be available on the original path, speed would rise, and thus the principal cost — journey time — would fall.

The curious thing is :

Knight's Paradox	When charging for access to a conduit, no more than the marginal cost of passage, no user is worse off, despite the price rise and revenue raised.

Those who "shift routes" experience the same speed on the slower path they otherwise would have encountered on the faster path, and are thus no worse off. Those who remain on the faster path see the price rise offset by an increase in speed when the others leave. Overall, there is one clear benefit: the income generated by the new levy. This, if properly invested, might yield improvement of the slower path, which might encourage yet earlier migration, and thus further increase speed and revenue, and so on.

Were the two services offered by different providers, you might be forgiven for thinking that one is very unlikely to encourage defection to the other. But this is exactly what vendors do in a free market. They raise their price, to the point that yields the maximum return. This is also the point that yields maximum *utility* — the best use of all competing resources.

Maximum income coincides with maximum communal value.

Downs took account only of people adapting existing journeys. We now must accept that they also both extend existing journeys — for example, by moving further away from their workplace, or taking a job further from home — and engage in entirely new ones, whenever additional capacity is provided. It is this tendency that strongly suggests that entirely new demand is stimulated by *any* increase in transport capacity, *whatever the mode.*

Incomes have risen and the cost of mobility has fallen. Demand may not be infinite, but it does clearly bear only a limited sensitivity to price.

Time would appear to be the only limiting factor.

Downs later picked up on this, and thus poured even more cold water on the idea of "relieving congestion" :

> "Some city planners have mistakenly concluded they could reduce surface congestion by building subways or other costly off-street transit systems. Any space these systems initially free up on surface streets, however, soon fills with more cars and trucks as ownership and use expand."

His remarks do seem to tally with experience. Indeed, we should expect no less from the market. Any space "freed up" on a mode previously saturated merely exposes additional value at the same price.

Demand will rise accordingly.

There are ultimately just two ways to match rising demand with constant supply: raise the price, or form a queue. If demand rises and price does not then expect a queue. Congestion is simply another word for queuing.

The problem is that price forms a very low proportion of the perceived cost of transport. To affect demand, it must rise dramatically.

Downs' Law has far-reaching consequences, even as it was originally put forward in 1962. Foremost, is that widening a congested conduit will offer only temporary respite. When generalized, it has one that is profound.

Simply put, *you cannot 'relieve' congestion.*

You can only use price to regulate demand, maximize the utility of all existing resources (conduits), and raise finance for new ones.

Modal inversion and the Downs-Thomson Paradox

Anyone who has ever run a business knows that some costs are *fixed* and some are *variable*. Where fixed costs are high, "economies of scale" may lead to an average cost that falls with each new customer. Where variable costs dominate, the average cost can often rise with demand, perhaps because some given resource is limited in supply. The fixed costs of building a 'wide' transport conduit may be high but space on it remains relatively plentiful. A 'narrow' one may be cheaper but space will be limited.

Suppose building a certain conduit was extremely expensive, but that enough custom has been attracted to cover the investment, though capacity is not fully exploited. More customers could be accommodated but cannot be enticed because their average cost remains too high, though the marginal cost of serving them may be low. It does not take much business acumen to see that revenue can be increased by offering newcomers a lower price.

Somehow, we must discriminate between 'early' customers, who essentially pay our fixed costs, and 'late' ones, who can push us further "into the black" afterwards. Some must pay more than others. Exactly *how* this can be done will depend upon circumstances. We must look for any discrimination possible in service provision. For example, it may be that freight can play the part of 'early' custom, and passengers the 'late'. Commuters may not like higher charges for peak-time service, but then it *does* have greater value to them. Shoppers might ride later and cheaper. A premium might also be set upon "first class" provision, beyond its additional cost, and so on. The point is that it can be done, and it *was* done with the railway.

A return may arrive slowly to begin with, when few people use the new asset, but as more customers arrive income builds. Marginal cost will fall as economies of scale take effect. An ability to carry many passengers at high speed suggests we name this the 'broadband' service.

Let us now think what might happen when a broad path, with its falling cost curve, finds itself in competition with a narrow path, whose marginal cost rises (Figure 2.11). This is how it was between road and rail.

If we begin with no one travelling and then ask what happens as people begin to arrive, it's pretty obvious that they will opt for the faster path, *i.e.* the narrow one. As their numbers grow, and they begin to delay each other, there comes a point where the two curves cross, and where it would have been cheaper to use the broader path.

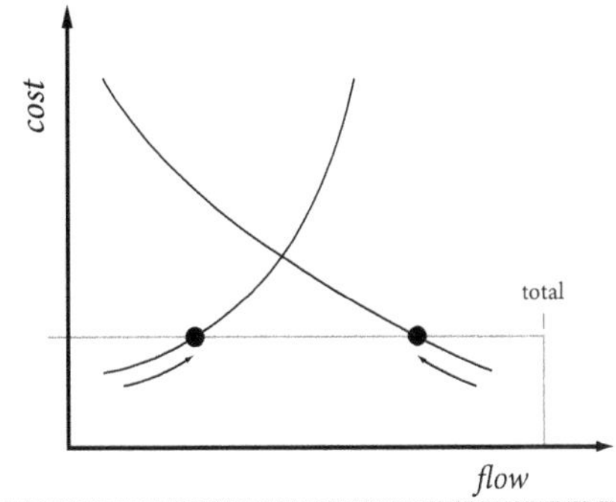

FIGURE 2.11 Competition between modes with rising and falling cost.

The trouble is that it will not be there. To be available, the broad path would need to have already served passengers in sufficient numbers to have brought its cost down to that point. But they all chose otherwise. Thus far, it has been uncompetitive. Even if its disadvantage was minute, *all* customers would still have chosen the alternative.

The broad path would never have been built in the first place.

Only two explanations are possible for the existence of competing 'broadband' transport. Perhaps the conduit was built before the narrow, and slightly faster, path became available. In this case, some of its users will undoubtedly defect. Migration will stop only when the cost of using each service becomes the same. A balance will be established.

An alternative explanation is intervention by a well-intending, but somewhat autocratic, authority. (A commercial operator would be unlikely ever to take the risk.) It might choose to build a broadband system (at our expense) and somehow either induce (via subsidy) or coerce us (via penalty) into migration to the extent that it becomes independently viable.

As we all know, such a system was, in fact, built first and was used by everyone. Its "operating point" (the black dot in Figure 2.11) moved a long way down the falling cost curve. The arrival of the car presented an opportunity for some to migrate and lower their cost of travel a little. As migration continued, the cost of using *either* service rose. If the cost of using the new mode rises more rapidly than that of the old, a stable equilibrium arises, but at a slightly higher cost than before.

Without interference, this situation would remain.

Now suppose that the balance of traffic on the two services is at a yet higher cost, though still below the point where the curves cross, and costs are equal. The narrow path will be congested. The broad one will also be unable to deliver its lowest cost. People will complain about both. In the 1970s, a man called Thomson asked how an authority might respond. In particular, he looked into the consequences of answering the popular call to invest in the congested conduit, in the light of Downs' Law.

The effect of such investment is to reduce the cost of using the narrow path and effect a slower increase of cost with flow. More, with shallow loyalty, will shift from the broad path onto the narrow. Although the reduction in cost may be slight, the effect can be profound, because a positive advantage in shifting continues and even grows as more migrate. Balance will now only be restored when the cost of using each service is once again the same. But that cost will be *far* higher, for *both* services — a disastrous result.

Such a catastrophe is called a "modal inversion" (Figure 2.12).

The understandably controversial idea that investing money on a popular transport conduit can actually makes things worse has become known as the "Lewis-Mogridge" position. It has been around since 1977, and is otherwise known as the

DOWNS-THOMSON PARADOX When a transport mode whose cost falls with usage is in competition with one whose cost is initially lower but rises to exceed that of the former, investment in the latter can result in a sharp decrease in the utility of both.

That railways exhibit a falling cost curve, and the problem it poses their operators, has been recognized almost since their inception.[30] Sadly, what can happen when rail competes with road transport was perceived too late.

We are now living with slower and more expensive transport as result, and the effect that has upon the economy.

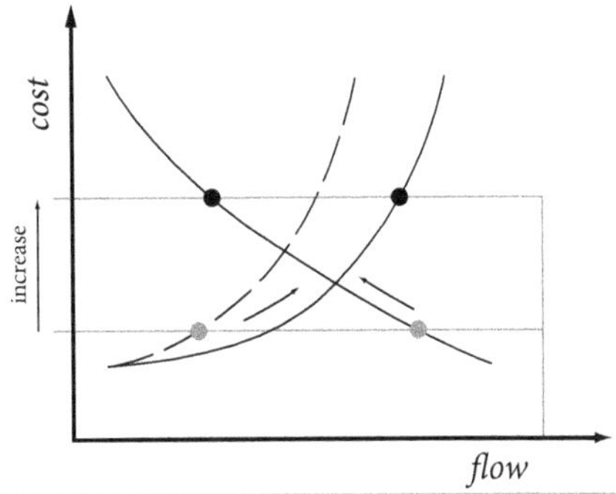

FIGURE 2.12 Modal inversion.

Modal inversion happened between road and rail in Britain in the 1950s. The result was not just a higher cost, however we travel, but that the congestible mode was to carry the *majority* of the traffic. It might seem less of a disaster given the distance that we now often live from a station. But in 1948, *every* British town of over 20,000 people was served directly, and 98% of those of over 5,000. Two in three villages were connected by rail.

The railway effected a more or less complete transport network.

Martin Mogridge, of the aforementioned 'position', is more properly credited with noting the inverse of the Downs-Thomson Paradox — that investment in the mode with falling cost can produce a disproportionate *benefit* to users of both services. We should emphasize 'can' because all will depend on the shape of those cost curves.

To reverse modal inversion would be extremely difficult, though it would appear to be an economic necessity. The potential payoff is the decrease in

30 Essentially, the problem is that track costs dominate and vary little with traffic type or volume. This encourages a higher charge to lighter traffic, whereupon it quickly becomes lighter still, and so on. Conditions form the antithesis of those required for growth.

cost achieved multiplied by the total number travelling. *Everyone* would benefit, whichever mode they choose.

Profitability varies out of all proportion with investment because not only would value (and thus price) to the customer rise, but *much* more business would be attracted from the competition. Railways might again be an attractive proposition for private enterprise.

Investment would be directed at reducing journey time, which is more likely to mean reopening closed lines and improving integration than raising actual speed over the track. Reducing delay in departure, connection, and journey completion, would be the priority.

The problem is that *focal* development has given way to sprawl, which is almost impossible to serve by any shared form of transport. Stations cannot be brought near enough doors without excessive cost, both in journey time and operation. It is cheaper by far to bring doors nearer a station.

Transport can only be addressed as an aspect of habitat design.

Paying the fare

However we travel, there remains the problem of finance, when initial investment is typically very high and revenue must be recovered over an extended period, via frequent small tolls. A similar situation lies in the movie business. Walt Disney once remarked how odd it was to have millions available, when each customer paid only nickels and dimes. The difference with transport is not the "nickel and dime" fares but that the investment needed is often *billions*, not mere millions, and that assets must typically pay for themselves over generations, not just years.

The capital cost of a new transport conduit does not vary as greatly with mode as some seem to assume. We have seen it emerge that a kilometre of motorway in Britain costs about the same as one of LGV in France or Spain (about £15m in 2005). We have also seen two differences emerge. First, the railway conveys at least *fifty times* as many people per hour, at around three times the speed. It thus offers more than a hundred times the value. Second, the marginal cost (of serving each new passenger) rises with flow on the road, whereas it falls on the railway. Without a viable railway, there can also be no proper choice in suburban, district, or regional, travel.

Recall that *price must vary with cost.* This applies to both road and rail. On the road, the purpose is to regulate demand. On the railway, it is to fill as many seats as possible, to make the most use of an expensive asset.

The railway exploited marginal pricing long before aviation. After freight financed its fixed costs (primarily track and signalling), operators quickly learned that the wealthy would pay handsomely to be carried around in luxury. Ordinary people were later to be carried in cattle trucks, until Parliament intervened, and compelled protection from the elements.

The aim on the railway has always been to *increase* demand.

Road pricing is needed to curtail it.

Understanding how to determine the value of a given path, and the consequent need to charge accordingly, goes back a long way.

The appalling state of the public roads in Britain, maintained by each Parish through which they passed, all but compelled their passing to private hands as far back as 1678, with the introduction of the 'turnpike' — a row of pikes (designed for 'turning' cavalry) — to prevent anyone bypassing the toll-gate. Not surprisingly, much debate ensued as to what the price ought to be, particularly when the path was popular and widening it was prohibitively expensive, as with, say, a bridge.

A French engineer — one Monsieur Dupuit — gained the credit for being the first to record (in 1844) how the amount folk are willing to pay to cross a narrow footbridge falls with traffic :

> "We want to know the utility of a footbridge which is being used free of charge at the rate of 2,080,000 crossings annually. Suppose that a toll of one centime would reduce the number by 330,000, that a toll of two centimes reduces it by another 294,000, and so on. We then say that for 330,000 crossings, the utility is about one centime, and that for the next 294,000 crossings the utility is about two centimes *etc.*"

No one has since put it more eloquently.

Of course, the variation is not quite as coarse as he cites. It would in fact be continuous. On the other hand, cash has to be collected in discrete amounts. At present, we are still stuck with the same rather primitive technology — coins — which presents a problem.

Cash has other disadvantages. It is far from convenient, in that it tends to be heavy. Like their paper counterparts, coins are now mere tokens, and possess the lowest practicable intrinsic value. This allows them to become smaller, lighter, and cheaper to make. However, we need very small denominations for toll-paying. Carrying many small tokens is a nuisance. Security is also a concern. Cash, by definition, requires no authentication. It belongs to the bearer, who need never verify ownership. We stand in need of something better, and not just for tolls. Many an electronic market now awaits a new form of cash.

Those little plastic cards have become a familiar way by which to pay, and an even easier way to borrow. If you prefer to avoid credit, you can also directly debit your bank account, *if* you feel happy to trust the vendor with access. A single card can provide both credit and debit services, but there is one big difference your bank does not wish to advertise. Should you choose credit, the bank pays, then you pay them. If the vendor overcharges, or charges twice, it's the bank's problem, not yours. They tend to win most arguments, being the supertanker to the sailboat.

Should you choose debit, you're on your own.

Banks are so big, rich, and powerful, that they don't even seem to much mind being the victims of fraud. Only after decades of such abuse, have they finally got around to making cards a bit more secure. "Chip and PIN" has only slowly taken over in Britain. There is now silicon in the plastic. Cards have acquired memory, the ability to calculate, and a means of getting information in and out (those little gold contacts). In other words, there is a computer inside.[31]

A few years ago, I had the privilege of working with people developing, and deploying, the ability to store cash in the card. This will be a revolution as big as when coins lost their intrinsic value, and a lot more noticeable. Cash will cease to be physical and become 'virtual'. An immediate application is obvious: it will become far easier to charge very small amounts frequently. For example, one might levy a tenth of a penny to display a page on a website. Silicon memory has now become so cheap that Bill Gates could keep his entire fortune in cash on a single card and spend it a cent of a cent at a time.

---

31 Smartcards have been made that even have both a display and keypad.

He'd better not lose it.

The card's processor is as important as its memory. It is this that affords a great degree of security. One can do much better than matching a four-digit personal identification number (PIN), which will no doubt please Bill. Experience has presented worse security problems, such as denying unwarranted access to a nuclear missile silo. Suitable methods are ready and waiting. There is even talk of scanning your retina at cashpoints (although this didn't stop THRUSH stealing a couple of atom bombs).

A card's ability to communicate is no longer limited to electrical contacts. They can use radio as well.

A whole new take on the turnpike has thus become possible. Collecting tolls can be fully automated. No one need even slow down. Charges can be adjusted according to as many criteria as you like, including time of day, or current demand for the path concerned. Price can now reflect the marginal, rather than just the average, cost.

If you *must* join a congested road, you can be charged accordingly.

The aforementioned M6 Toll Road is interesting for another reason. As its name indicates, it is a toll road, built by a private operator, and paid for by subscription. After a year of operation, it was deemed a thorough success and gives the lie to the notion that roads, especially trunk roads, must be the gift of the State. The need for public regulation may remain, but the need for public provision has vanished.

Experience with Britain's first toll road in modern times exposes the problem with a piecemeal approach to road-use charging. After the first year of operation, its masters were able to report genuine relief of the existing parallel public road, and without congestion on the new one, which could perhaps be relieved anyway, simply by increasing the 8p/km charge.

The 'but' followed :

> "However the M6 north and south of the tie-ins of the M6 Toll showed traffic growth above the national average."

Any communication network is holistic in nature. It cannot be divided up into individual links or partitions without causing effects across the boundaries introduced. Close a single station on the railway network and you remove a destination from each of the remaining 2,499 stations. This is called a 'combinatoric' effect. Similarly, if you create a shortcut, like the M6 Toll Road, on the road network, you affect the choice of routes available for *all* journeys across it. In short, you can expect congestion leading up to the junctions between the old congested link and the new uncongested one.

This is but another example of Downs' Law.

Of course, as has been noted by the road lobby, such roads break the commitment made by the UK Government that road-use charging will be "tax neutral". In any case, if it *is* accompanied by a reduction in fuel tax, it will have no effect upon choice of mode. There will be no corresponding switch from road to rail, from which the only possible relief from road congestion can ultimately come.

Even were it not tax neutral, and instead deliberately added significantly to the cost of motoring, road-use charging is most unlikely to bring about a reversal of the modal inversion between road and rail. Given that the cost in time forms two thirds of the total cost of use, even when *all* the running costs of the car are counted, *doubling* those running costs will add little more than a third to the full cost of driving.

This would mean charging *five times* the toll on the M6.

Clearly, no politically acceptable rate can reverse modal inversion.

Only shorter journey times by rail can possibly accomplish that.

About the most common objection to tolls, besides actually having to pay them, are the loss of privacy and the issue of *trust*. We are just getting used to our PCs dialing their corporate parents without telling us what they are up to. When they use radio, there isn't even a plug to pull. Once cash goes electronic, *every* financial transaction can be traced. When this extends to paying tolls, our movements can be watched in the smallest detail.

This can even extend down to our movements around a building. I was a very early installer of a card entry system and sought experience with such things. I came across an anecdote of someone using one to account for a robbery. They were not only able to isolate the thief, but could also trace their movement through the building the entire day. Each time they entered a room, they had to swipe their 'enter-card', and entry was centrally recorded. With radio-equipped cards, even finer detail can be added to records, yet users will be aware of none of it. They no longer need to swipe. They will forget they are being watched.

On the road, we may keep this in mind, but the demand for trust goes further. If the basis on which the toll is calculated is complicated then it is difficult to verify. The system is then open to corruption and fraud. Furthermore, charging according to a factor that is variable, such as current demand, can conflict with the freedom of the customer to make an informed choice. By the time you arrive at the toll-gate, the price may have changed. The problem grows worse when planning a journey along multiple connecting paths. You can only *trust* the provider.

Curiously, the same objections apply to using a mobile phone, especially in a foreign land. As I write, a terrorist was traced across Europe by his use of a phone, and then arrested. He was clearly unaware of the possibility. I doubt if he had much idea what he was paying for the calls either. These difficulties have not slowed the growth of the mobile phone. Such precedent suggests we are happy to accept the risks, given a new freedom — in this case, to communicate wherever we are, whatever we are doing.

Liberty has always depended on accountable regulation. And there are always new challenges. In any case, there are alternatives. When cash goes virtual, it need not necessarily lose its anonymity.

One proposed solution for the privacy problem is *prepayment*. Effectively, a particular form of cash is loaded onto the card before travelling. Some of it is simply deducted on passing the toll-gate, which need know nothing of *who* is passing. Less trust is required, since the card knows who took payment and how much, and can either tell you or verify it was the correct amount itself. This is a rather nice example of how computers can act autonomously, like a fairly capable human assistant.

You may not trust the State, or the service provider, but you might trust your very own intelligent electronic purse.

All transport amenities, regardless of mode, need investment. Everyone who benefits can be called upon to make a fair contribution. Good transport adds to the value of both property and business. New development can be made subject to "planning gain", whereby permission to build is granted subject to a contribution being made towards establishing amenities, including transport. Developers in Britain are increasingly asked to help build libraries and schools. They might also help with pavements, cycle-ways, roads, and stations, to the extent they can recover when they rent, lease, or sell what they build. On occasion, they even volunteer funds.

Planning gain causes little distress to residents and entrepreneurs, if only because they are more conscious of paying for a home or office than for the means of going to and from it. They are less likely to celebrate paying for that which already exists, and which they may not themselves use. Hence, planning gain is a more attractive opportunity for funding *new* infrastructure, on which development is to be predicated.

A common argument is that people who live nearby benefit, both from the amenity itself and from the higher property value it confers. Hence, it is said, they should pay some sort of "property tax". This is much less easy to sustain. For one thing, they will already have paid the increase in property value when they bought their house or business, if the amenity was already there. If it was not then the opportunity to charge for it was missed. In any case, it is not usually considered fair to charge someone retrospectively, when they cannot decide to go elsewhere instead.

On the other hand, should the local transport infrastructure fall into ruin, both business and residence will surely *lose* value. There is thus surely a case for paying something towards its upkeep, even if resident and proprietor seldom use it themselves.

An accountable local and regional authority is clearly warranted.

Given the technical ability to charge for travel according to cost, whatever the mode, and to care for the common good via regulation, it is hard to see any reason why any public authority should act as provider. Given also that doing so carries considerable liability and no profit of any kind, it is similarly hard to see why any volunteer so to do.

The need for social inclusion has often been cited as a justification for the public provision of transport, but this can surely be achieved through mere regulation, which requires the raising of no taxes.

There are three problems with public provision. First, any subsidy lies outside of market forces. The price of the subsidized service will no longer be subject to competition. A result may well be that the *total* price paid — direct charge plus subsidy — will be greater than that of a competing service. When the perceived (directly charged) price is lower, the dearer one may gain advantage. It will then be free to become less efficient, without risk of losing its market — a trend leading to ever higher subsidy.

Second, a public authority is likely to act only over the short term in a way that yields the most votes on election day. On a road, peak capacity arrives only after speed has fallen. As we have seen, this infers low value and a poor return on investment. Rather than apply unpopular access charges, to regulate demand and allow investment in rail, congestion may well be tolerated. Voters are left with jammed roads and no alternative.

To a private provider, things look very different. Their interest will be served by achieving maximum value, over the longest possible term.

When people last got about by rail and bike, there was little exclusivity on social grounds. Both bike and rail ticket were affordable to all but a few. The need to travel more than a mile or two was comparatively rare. By and large, access was equitable, so there was no real case for public intervention, via either regulation or provision. It may be different now.

Social exclusion may actually have been manufactured by the modal inversion between road and rail, which has greatly increased the cost of travel in both time and money. Its effect has been exacerbated by the interference between automobile and bike — the cheapest mode after walking — and the proliferation of sprawl, which has forced more and longer journeys, and made the use of a car obligatory.

Because of these changes, there are now many who need help, even to make essential trips. Not only has the need for more assistance been created, but answering it has become very expensive and difficult. You cannot regulate access to the private car. Its very privacy denies this. But with so many journeys now possible *only* by road, public provision of some other option becomes mandatory.

In Britain, as in many other countries, bus services receive massive subsidy, typically in meeting their track cost — usually hidden under road building and maintenance. However, we have seen that the bus is often more expensive to use than a private car, even *after* heavy public subsidy. It is also the slowest form of mechanized transport ever invented, and contributes significantly to both local pollution and congestion. In many town and city centres, it is the dominant source of both, causing incalculable damage to the health of both people and the local economy.

Sprawl cannot be adequately served even by the largest fleet of buses with the greatest subsidy. The diversity of paths will always suffer a "combinatoric explosion", leaving only a tiny proportion of demand met.

Social security is already meeting the cost of a private car for some claimants, simply because no affordable alternative exists for them.

The cost of reversing modal inversion is undoubtedly astronomic.

But it may be cheaper to meet it, in the end.

3

A few short stories

Each year in Britain, around six out of every hundred thousand of us die in a road-traffic accident (RTA). In the USA, it is almost fifteen.

As a rule, ten times as many are seriously injured.

Many, many, more go on to be emotionally scarred for life.

Obviously, the annual carnage it brings about is a very significant cost of the car. To make reference only to numbers would be to fail utterly to count this cost. Statistics are as dead as they are dull. While it would be out of keeping in any scholarly work, and even here it may offend certain (particularly Anglo-Saxon) sensibilities, to include what is inevitably highly emotive material, it is nonetheless *essential* to expose the reality.

No one is fit to judge what is an 'acceptable' casualty rate without some awareness of its meaning.

To bring any real incident to life would necessarily intrude on the privacy of both victim and perpetrator. A narrator would also be compelled to tread the minefield of political and judicial correctness. Instead, we shall follow the paths of fictional characters, living imaginary lives, but who suffer an accident endured, along with all its consequences, by very real people.

Any similarity between a character described and any real person, alive or dead, is purely coincidental.

3.1 Just turning left

They met through a friend, a long time ago. Sandy was a little on the wild side, compared with Helen, and had "got into" live music. Let's face it, the sax is a sexy instrument. It held her attention a little longer than a bloke called Dave, who happened to be behind it that night. Well, they got talking, and, as they say, one thing led to another. She was just sixteen at the time, but then he was only nineteen, so what the heck.

A band attracts a crowd, which is a good job, since an audience is always a good thing. Friends do a good job of getting the message out and bringing one in. A lot of fun was had, but the time came when her A Level results arrived. She'd got a place at Oxford. There's no way she could face down her folks and stay home. Anyway, he'd never let her do that. He was just an electrician, but he'd take his chances. And she'd take hers, though not without a few tears, now and then.

Oxford was not so far away, and her friends made it quite easy for both of them. Snobbery is less common there than some might think. She came home regularly too, for her folks as well as him. Though occasional temptations past by on both sides, they were still together at the end.

A university of that calibre begets many professional jobs and could answer her qualifications much better than her old home town. As is far from unusual, she had fallen in love with the ancient city — not with the spires or the ivory, but with its diversity and character, and the colleges, bookshops, libraries, and cafés, that were its vital organs. Somehow, it defied modernity; ignored it even, with casual indifference.

She found this reassuring.

A degree in English Literature is hardly preparation for industrial production, but a First from Oxford in anything is marketable. Publishing still thrived there, despite the onslaught of electronics and aggressive competition from what remained known locally as the "New World". An offer of a post coordinating marketing and sales for a large 'established' company was gratefully accepted, but only after talking to Dave.

He was not going to argue. The social life in a large university town was attractive enough anyway, but the clincher was the chance to get a place of their own, together. Commuting would be a pain and money would be tight, but they'd manage somehow.

In fact, they managed rather well. Within a year, he'd got a job in Abingdon, just twenty minutes drive, instead of forty, and with prospects that paid off. He had always been bright, and a quick study. This time, they noticed and started giving him training and more responsibility. Interesting, satisfying work too it was, making medical instrumentation. It was a buzz to know what you did might save life or limb.

She too began to climb the ladder. Before long they found they had to keep a bit quiet about their holidays when visiting her sister Sally, who now had a young family in tow, and had thus surrendered one income for new outgoings. "Alright for you DINKS," she'd say, "dual income, no kids." But there was no animosity. Neither sister had any understanding of the open warfare they heard of elsewhere. They had always been close. Helen found herself joining in the games, but still stayed well clear of the nappies.

"Do you want those," said a nervous Dave, as he began restoring order after a particularly rumpustuous visit by his in-laws, with their two offspring, now nuclear-powered toddlers. A pause followed, then "Naw, don't think so. We're alright as we are, aren't we?"

"Don't ask me! You're the one with all those hormones and a womb," came the reply.

"Perhaps we should get a dog," she said. "I don't think we could look after one, though." Another pause. "Do you want kids?"

"Some day maybe," he offered.

No more was said for another few years, until Sally had another, this time 'unintended', but nonetheless welcome, addition to her family. Now thirty-five, she told tales of lists thrust in front of her warning of the possible dire consequences of having a baby so late. "Downs' alone goes up to one in a hundred by the time you're forty," she said, after an 'amnio' dismissed a negative blood test. She had the sense to say no more, though her sister was still only twenty-nine.

It suddenly occurred to Helen that the intent was there. She'd just been pushing it to the back of her mind, and telling herself there was plenty of time.

What if Dave didn't want the same thing?

She told him what Sally had said that night, when he came back after midnight from a gig in London, complaining he was "getting a bit old for it". Childhood sweethearts keep nothing from each other.

"Well, do you want children or don't you?," he said. "Maybe it's time to decide."

Again that pause, that he'd come to know so well.

"Yes," she eventually replied.

He was not really surprised. He simply hadn't known what she wanted, until then. He believed her. "Well, we'd better get started! In the morning though. I'm knackered right now. Better chuck those pills away."

"What about you," she said quietly.

This time the delay was his, which worried her a bit. She had never known him uncertain.

"This might not make much sense," he said. "I had a dream, the other night. You know, the kind that's real. It woke me up."

"Go on," she said, after a full minute's silence.

"Well, there was this girl."

"Oh," she said, tacitly.

"No, no. A *little* girl. Standing in the doorway. Over there. I knew her, somehow, close like. Very close. About five, I suppose. She didn't say anything. She didn't have to. It was there in how she looked at me. 'Why didn't you want me Daddy?' "

She hadn't expected this. It was the first time she'd ever seen a tear in his eye. Apart from when he lost his Dad that is, to lung cancer.

"We'd better hope for a girl then," she said.

But he was already asleep.

Reality has a way of catching up, even with DINKs, basking in industrial prosperity, and post-industrial comfort. Two years had passed, and there could be no denying something was wrong. Doctors, and hospital departments, became part of life, as did the gray malaise that settled into the background of each waking moment.

A sense of failure is always the harder, after nothing but success.

"I'm *not* going for any more bloody tests," she cried, after his hint that there was nowt wrong with *him*. "It just makes it worse. Bloody doctors. They'll never admit they just don't know. They don't want to lose their sense of superiority. Smug gits, the lot of 'em."

He sank a little lower. He'd never heard her resort to swearing and insults before. One of the things he now knew he loved in her was her inability to get flustered, or even to understand those who did. He remembered arriving in Athens, just an hour or two before sundown, with nowhere to stay. With no Greek, and very little cash, he was getting wound up. Having to defend an attractive and exuberant young woman, with not the slightest perception of risk, in a strange city after dark, was not his idea of a holiday.

She found a place within minutes. The cockroaches had to be seen to be believed, but it was fine.

And now he heard her yell "If you're so perfect, why don't you find somebody else."

"Maybe I will," he reacted, instinctively. But she noticed the tear in his eye, and fell silent. "No you won't" she said quietly, taking his hand and kissing him gently. A little later, they snuggled up and watched a movie together, with a beer and nibbles.

That night, he awoke to her sobbing. At his touch, she turned and cried herself to sleep on his shoulder.

After another two years, they were beginning to accept the situation, and to make the most of what they had, in each other. A deeper sense of well-being came to stay. Which made the shock all the more profound when she discovered she was pregnant. She chose her moment well. He just gazed at her, with disbelief at first, but then a growing sense of bond and pride. As couples had found down all the millennia of history, there was always more to explore in each other and in what you become together.

They knew there was a long way to go, and that it would be far from easy. Thanks to Sally, they both knew that the age of indulgence was drawing to an end, and one of service due.

Her Mum and Sally were over the moon. She was to join a timeless club, and sow her stitch.

Plans were laid and the house rearranged; a holiday cancelled, in favour of one that called for no jabs and less expense. Her workmates were wonderful and her boss could not have done more. Her job and her position would remain secure, even should she take more than a year's maternity leave. But then, uncommonly, her boss was a mother too, and understood the dilemma a woman faced, mixing career and family.

Another few weeks, and she'd close down her desk.

Another Monday morning. Up at six-thirty, breakfast at seven. Some holiday things to sort out in town at lunch-time; to be straightened out with Dave first. It was gray outside, but she felt sunny. He snuck up behind her and hugged her so sensually that for a moment she was tempted. No time. She teased him a little, then sat him on his stool. The toaster popped.

They kissed at the gate and he told her to take care, as he always did, as he headed for his car.

She walked the bike across at the lights, onto the green cycle-lane opposite, and set off, thoughts of the day's work ahead coming to mind. Though it's not enclosed, the bike gave her a little personal space, and time to think a bit. Riding came as naturally to her as walking.

She'd been cycling since she was six.

Forming the westerly entrance to the city, The Botley Road was as busy as ever, but still flowing this early in the morning. For a spell, where width permitted, the cycle lane dodged off-road, and up next to the footway. But then it became treacherous, as the road ran through a chicane or two; the worst being the one under the railway bridge. Old Brunel had to consider only the horse and cart, so the road dipped beneath the railway through a tunnel not wide enough for cycle lanes. You either commanded the space you needed, or got off and walked along the narrow pavement.

As she prepared to negotiate the tunnel, she became aware of a large building-supplies lorry drawing alongside. Nothing unusual. On a bike you get used to rubbing shoulders with metal giants. But it suddenly swerved inward to enter a service road just before the tunnel — Roger Dudman Way. She had no idea who Roger Dudman had been, but it was a name she would remember too well. There was no time to turn inward herself, or even to lay the bike down. Eighteen tonnes of steel hit her, broadside, knocking her to the ground. The nearside rear wheels ran straight over her chest, smashing nine ribs.

Her baby died immediately from the impact.

She had no breath to scream. Others did it for her, when they saw the bloody horror emerging behind the vehicle. One kept her head and called for an ambulance, wondering later just how many have been saved by a mobile phone. It took fifteen minutes to travel five kilometres across town, through the traffic. Somehow, she stayed alive, and even conscious, right up to her arrival at hospital.

It was just another day in the life of heroic Accident & Emergency staff. Perhaps professionalism can be expected, but not the shining tenderness among all who cared for her. They found a broken shoulder, lacerated liver, and damaged pancreas, but worst was the state of her lungs, one of which was punctured.

Lola, it was, who was to go through her things and find out who she was, and who would want to know what had happened. Helen's bag was opened, and her wallet. Even this took time and expertise. Finally the phone rang at her parents' home, and a little later, Dave's mobile sang its merry chirp.

He cursed the interruption before he answered, and the blood drained from his face.

Like extras from a TV soap opera, they were gathered in a room the hospital, with systematic efficiency, sets aside for such sadly frequent occasions. Someone brought coffee, as they struggled to cope with the reality of what was happening.

After an interminable period, which might have been minutes or perhaps hours, a woman with a blue-white medical coat appeared. She was just a little older than Helen. She detailed what the ambulance crew had picked up from witnesses, and particularly one Dolly Weston, who had called them, and who had even given such First Aid as was possible. A Doctor Atkins then took them through the catalogue of injuries. Cruel experience taught doctors that raw facts were the easiest way in the end. But one could never resist eyes that pleaded for prognosis.

No one ever accepts "I don't know" from a doctor.

"I'm afraid it's the damage to her lungs. You must expect her to go downhill, I'm afraid."

"Die, you mean," said Dave, who could no longer focus, in any sense. "I'm sorry," he added.

"Can we see her? Is she conscious?" her father asked.

"Just from the doorway, for now" replied Atkins. "She was conscious when she came in, and she might be able to cope tomorrow. We can manage pain pretty well nowadays."

She noticed Dave staring at her, now with a stillness and intensity.

"The baby," he stuttered. "She was seventeen weeks..."

Even the doctor, a veteran of many lost battles, and a witness to many a tragedy, visibly choked a little, as she replied softly "I'm sorry."

Evolution has earned us the "blitz mentality" — the ability to recover and function even within the deepest trauma.

"Will we be able to bury him, at least."

"Yes," came the clinical response.

After a while, her father added "May we see her now?"

Dolly Weston, first on the scene, and calm enough to act at the time, had to be taken home from work. She had suddenly begun shivering uncontrollably, and had become unable to speak more than a word or two at a time. She would go on to suffer nightmares, depression, anxiety, and an emotional detachment that would have been unthinkable before.

Hugh Jones, the driver of the lorry, and second on the scene, would also suffer. Having always thought of himself as hard, he threw up on seeing Helen lying in the road, and became completely dysfunctional. He would now become easily and severely irritated, and developed a tick, on occasions. Wild mood swings would assail him, and he began to avoid emotional contact, and shunned his old mates down the pub.

While he could find no fault with his driving, he blamed himself. He had a daughter of not dissimilar age, and kept seeing her in Helen's place. Sometimes, he would descend into helpless tears, though he could usually manage to postpone it long enough to reach privacy.

His life would never be the same.

The court did not fully exonerate him. He got a fine and seven points on his licence, though they acknowledged there was nothing different he should have done. A bicycle in an adjacent cycle-lane is simply invisible to the driver of a heavy lorry. Mirrors are located to assist in avoiding other heavy lorries. Even cars only become apparent at the rear.

The design fault has long been acknowledged, but never remedied.

It's the same with every bus.

After three days in Intensive Care, Helen had recovered enough to be moved to an ordinary ward, denying the initial dire prognosis. Having thought they had lost her, Dave and her folks were over the moon to learn she would live. Dave even risked a joke, and promised to buy her a new bike — one at least ten feet high, with flashing lights and a radar reflector.

He'd been intercepted on his first visit, after she regained consciousness properly. Estelle Atkins took him back to that literally dreadful room and said simply "Don't raise the issue of children yet. We've told her about her baby. I'm sorry to tell you that she can't have another. You must understand how lucky she is to be alive, and how lucky you are to get her back."

It was a message timed with skill, learned with bitter experience. She knew he would not yet have found the time, never mind the courage, to think of trying for children again. A hope not yet imagined cannot then be lost. Also the sheer joy of seeing the love of his life live again would carry him through.

There was one thing he had already decided he *would* mention, just because it emphasized their good fortune.

After the best attempt possible at a hug, under the still heavily plastered circumstances, it was she who spoke first, wanting to know how her folks were bearing up. He could tell her that her parents were on standby to take care of her in her own home while Dave was at work, if that was what she wanted. "I think they may need a bit of that," he said. "Sally's hot on my heels, you know. Careful to defer to me but she *really* wants in."

She came over thoughtful, and looked out the window, then down on the bed. "There's something that's a bit odd. I keep feeling like I've been through this before. I've never had, what's it called..."

"Deja vu?" he offered.

"I have to tell you something," he began, trying not to sound too ominous. "If I know, then you have to too, I think. It's... well, it's happened before. More or less the exact same thing, on the same spot. To someone else... obviously."

"When?" she asked.

"About four years ago. A girl called Sandra... Sandra Meadows."

Helen glared at him fixedly. "Was she OK?"

"No. She died. And that's just the worst one. If I'd known how many cyclists get hit on that road, I'd have been riding shotgun."

He began to crack. Of a sudden, she realized what he'd been through, and reached out for him. Long ago, she'd learned of the small boy inside every man, no matter how big and tough.

It was left to her Mum to tell her about her extinguished fertility.

Mothers, after all, must cope with everything.

After ten days, they shooed her home, though an ambulance was necessary. It would be a while before she could fit in a car again. Dave had agreed to let her folks help her in and settle. It was only fair they should get the time with her. Caring is what parents do. And they knew well enough to leave her alone with a cup of tea on her patio for a bit.

No anger or bitterness or blame polluted her mind.

She just mulled the madness.

In her eyes, the house had changed. She could see where a child might play — in the living room, and out in the garden. She found herself still believing it would happen, even feeling excitement, before the truth overwhelmed her. No mischief would disturb the quiet now.

Strange how peace can fail in silence.

Postscript

In the year previous to "Helen's" experience, precisely the same thing happened to another young woman at a different junction in Oxford. Her body was similarly crushed. The same hospital staff saved her life, after fifty hours of surgery and a dozen operations. Within three months, the same thing happened yet again, twice, on the very same spot.

On both these occasions, it was a bus that ran over the cyclist.

3.2 A slight distraction

"You really going to stay here?" asked Dave with no sneer, but more than a little derision.

"So what's up with dear old England?" said his friend Stuart.

"Like one of Dad's old records... how does it go... '*too old and cold and settled in its ways here*,'" he sang in reply.

"I think she was talking about Paris," chimed in Diane, Stuart's girlfriend of four years now, an aficionado of sixties folk.

"Who?" said Dave, with typical prevarication.

"Joni Mitchell," said Diane, standing her ground, "as you well know."

"You're gonna marry this one?" came Dave's retort, as he locked eyeballs with Stuart. Diane looked away embarrassed.

"One fine day," he said, smiling at her.

They were at their favoured curry house, of the twenty-three now populating, if not defining, the Cowley Road. It was Saturday night, and they had set aside the textbooks for twenty-four hours. Medicine was pretty demanding, and they both wondered if they would really make it through such a gruelling six-year marathon. But like many before them, they had agreed it was too late now. There was too much already invested. It was doctorhood or death. Anyway, what else would complete them, now.

"Seriously, though," continued Dave, acting as though he'd been rudely interrupted, "it's a big planet. There are plenty of places with a better climate, better pay, and more going on. You can't plan anything here, in case it rains. Anyway, everything's broke and nobody's fixing it. You really want to be a prisoner to the NHS the rest of your life? Not me mate. The day I get my ticket, I'm on the plane... assuming the airports still work."

"I think it's more a matter of whether you can get there," agreed Stuart, having recently been stuck in York for the night, at his own expense, after the 'failure' of a train.

"I thought about that," entered Tracy, Dave's latest concubine, every bit as lovely as the last.

How does he find so many women who act as good as they look, thought Stuart. She was three years his junior and reading Modern History at Somerville, though she could have been from any class. It really was getting hard to tell, now it had become so 'uncool' to "talk posh".

"Process of elimination," she continued. "First boot out dictatorships, large and small."

"Whoops, there goes China," quipped Diane.

"Then all the places at war, or where *absolutely* nothing works. Then all those places where they'll now want to blow us up... after Blair that is... and you're left with about half a dozen countries."

She might just last another week, thought Stuart.

"I've always fancied Sri Lanka," he added.

"War," came the synchronized reply.

"There's all of Western Europe. Maybe Eastern Europe too, now," chimed the redoubtable Dave.

"You'd learn the language... adopt the culture?" queried Tracy, genuinely curious.

"Yeah, maybe," he replied. "Spain's got a brand new NHS, properly funded. Why shouldn't I heal a few sick Spaniards? I love Spanish culture anyway. And the beer's good."

"Yes, but would you adapt to it?"

"What, the beer?"

"The culture, fool," she dug her fingers in his ribs. He responded by tickling her.

Dave could always precipitate silence, given that he generated most of the noise. After a moment, he took a more serious air and announced, "I've already decided. It has to be the States. Where can you do better than in Rome."

"You plan to live among Americans?" said Stuart, aghast.

"At least they can enjoy their national sport without going ape," countered Dave, dryly. "They can still take their kids to a game."

"I think it was something my Dad told me Thatcher said," he added a moment later. "Someone asked her if Britain should have a space programme, and she said 'Oh no. We'll leave that to the Americans.' We leave everything to the Americans these days. The good stuff and the bad. Nothing ever happens *here* anymore."

"So when did you decide you wanted to become an astronaut?" said Tracy, sealing her fate. Dave did not take well to mockery.

"Speaking of flying," he escaped, looking at Stuart. "You want to come gliding next weekend?"

"Ok. I'll try anything once. Where?"

"Weston. It's only just the other side of town. They said if we come late afternoon, it's quiet, even on a Saturday. They take you up twice for twenty-five quid. I'll pay."

"No, I think I can find that. Yeah, sounds good."

It was another routine day for Adele. She collected her things and left the primary school, where she worked as a teaching assistant, and set off in her car to collect her thirteen-year-old, Austin. Things seemed stale, now they had grown used to life without Michael. It must be all of two years, she thought, since her man had left. She was still quite unable to understand how someone could just cut loose like that, when nothing really seemed that wrong. But she had no choice now but to accept that he simply didn't want her anymore. It was harder to accept that he didn't want his own son either. It was harder still for Austin — visibly so.

She had to wait longer and longer outside his school before he would appear. Some poor lad was being beaten up in the playground, but no one seemed to care. Adele cared, but didn't dare intervene. Someone tried two years ago, and ended up in the Infirmary with a serious knife wound. It broke her heart that nothing was done, although she'd admit it was far from easy. Some of the boys were way beyond redemption by fifteen, and many others followed, as sycophantic acolytes, if only for their own protection.

Austin was fairly tough, and embittered enough now himself to follow, though nowhere near vicious enough to lead. Guiltily, she admitted to herself that he stood more chance as a perpetrator than as a victim. She may have lacked the spirit to rally intervention, but she had heart enough to feel for both victims and their parents.

Sullenly, Austin finally appeared. Without speaking, he sat down heavily in the back. She knew better than to ask about his day, and better still than to ask of any scholarship undertaken. Reflecting on her own *happy* childhood, and all the encouragement her folks had given, she could only wonder yet again what on earth had gone wrong.

On arriving back at their little flat in Headington, Austin headed straight for his room, slamming the door. She collected the post, groaning at the bill, lost in the junk mail. It had to be faced. Over a cup of tea and a ginger-nut — the last of her indulgences — she opened it and laid it on the table alongside the notes she had made on her finances. Mike had left her with no savings, and sent only sporadic and diminishing quantities of cash.

Having never really concerned herself with money, she did her best. Most of the bills she could cope with, but the Council Tax defeated her. First, it was by far the biggest. Second, for some obscure reason it would stop for two months of the year, and then restart, making budgeting all the harder.

Already overdrawn, and with growing credit-card debt, they were steadily sinking.

She stared out of the window, determined to enjoy the biscuit and tea, but her mind would not allow it. Something cracked a little, and a tear despoiled the gas bill, across which a triumphant 'paid' had been written, just last week. It jolted her upright again. Tea to cook. "Keep buggering on!" she recited. If Churchill could, in the midst of a war, then so could she, resolving to call Mike the following morning. Familiar with his habits, she knew she could catch him early on Saturday, and in a good mood. At least he might send something towards his son's birthday. He could not put it off. It was just a week now.

The fact that she tried the same last year and got nothing more than a broken promise did not deter her.

A week later, Dave was cooking Saturday lunch, while Stuart did his best to help. "You'll make somebody a wonderful wife one day," said Stuart, with a poor imitation of a Yorkshire accent.

"So Di's happy to skivvy for you is she?" he retaliated. "Knowing your cooking, I think she'll have to," he added.

Stuart capitulated, focusing on the green peppers.

"Are you sure... I mean... are you really sure you're right for each other? That you'll still get on ten years from now?"

"Why not?" came the hesitant reply.

Dave gave up and tried a different tack. "Don't you ever get tempted, you know, to try other pastures."

"I suppose so, but... I just can't treat women that way. Anyway, I'm happy with Di. Suppose I'm in love."

This defeated Dave. He hadn't really considered the possibility. In any case, he didn't really know what it meant, since, he had to admit, he just hadn't been *that* smitten yet. He smiled, relishing the thought of something important he hadn't yet tried, and feeling sure it would be a bit more dramatic when he did.

His smile dropped a little; why did they call it 'falling' in love?

That didn't sound so good.

The cooking smells penetrated and thoroughly distracted him. He loved food. It had to be taken seriously, relished, ...how could he put it... *invested* in. Yeah, that would do. Stuart was quite happy with a soggy sandwich, or whatever was left over from last time he summoned up the will to cook. To Dave, this was sacrilege, in need of correction; but he had a selfish motive too. Apart from an evangelical urge to save his friend, he knew he needed to understand low aspirations. That few felt the need to aspire in every last one of life's alleyways was simply beyond him.

"No. There's nothing else for it," he said at last, to a Stuart thoroughly preoccupied with a cucumber. "Diane and I will simply have to elope and leave you to Tracy. You know she fancies you..."

He was cut off in mid-sentence by a flying lettuce.

At the same time, across town, Adele was busy. Mike had actually come through. A cheque for a hundred quid had arrived two days before, along with a simple note, saying "Give him a good time." When she told Austin, the change was immediate and striking. He smiled at her, saying "Really?"

She showed him the note, and asked "Do you want a party?"

"Can we go to Drydens? Where Steve had his."

"OK, but we've got to make this cover it alright? You can invite six. No more, understand?" It was then only Thursday morning. There would just be time for invites.

Now it was Saturday lunchtime and all the modern problems of parental logistics were descending, including the timeless one of keeping order. Austin had insisted on bringing along one or two with the simplest of "learning difficulties". They were completely out of control. Like the school, she must face the fact that they were someone else's kids, and there was little she could do. Simply getting there was her main worry. The reservation was for six and it was now five-thirty with no sign of the fellow parent, who was supposed to help run seven unruly teenaged boys across town.

So much rode on this party. Adele could not bear for it all to go 'pear-shaped'.

She made her decision, and began to herd them all into her own car — an ordinary, and elderly, family saloon. A quieter and reassuringly well-mannered lad took the front seat, fastening his seat-belt before even being asked. The rest revelled in the chaos of the back seat. Later, she would not remember whether she had made any attempt to persuade the boys to use the three seat-belts there.

They set off in a hurry, heading for the ring-road and south.

As they trooped off in the car, Stuart began to wonder what he'd let himself in for. "Tell me again, what we're doing. Are we meant to fly the thing ourselves today?"

"Don't worry," calmed Dave. "The instructor's with you the whole time. They do the take-off and landing. You get to take the controls just to fly around a bit."

"How do they get you up there?" worried Stuart.

"There's a winch. You go from nought to sixty in about two seconds. It's great. Anyway, haven't you ever dreamed of flying?"

"No," came the truly unexpected reply. Dave glanced across at his friend. He thought everyone dreamed of flying.

"Can we go a bit slower today, please," said Stuart, knowing it was pointless, as they roared up the northern carriageway of the ring-road, in the outside lane.

"What the — is he *doing*?" was the last thing Dave ever said, as the gray saloon loped and slid across the ten metre green central reservation. It seemed to happen slowly. Stuart remembered it in slow motion, as though his mind had been recording video at several times the normal frame rate, like they do for wildlife films from time to time. The oncoming car appeared to rapidly accelerate, but it was just perspective.

There was nowhere to go. The road was crowded in both lanes. In the last split second, he tried to swerve left, as he braked. This saved Stuart's life, but destroyed any chance Dave might have had, as his corner of their car took the full brunt of the impact. The bang was the loudest noise Stuart had ever heard. It was followed by the sounds of hell itself, and one more bang as something hit their rear, whipping back his neck.

An image that was to stay with him was of a crystal shower of shattered glass hanging in mid-air, frozen in time, like a scene from *The Matrix*.

Eventually, the sounds of metal being tortured and torn passed into silence. A light breeze ruffled nearby trees. Stuart found himself able to release his seat-belt and even open his door. Their vehicle was still upright, though leaning forward and to the right a little. Its offside front was gone. Fearing fire, his instinct to flee was overwhelming, but he forced himself to look to his right. A moment was enough to tell him there was no Dave anymore. He turned the other way and vomited, painfully. His flight instinct then took control and he became blind to the agony in his chest. He burst out, fell, crawled, and then staggered away, finding himself in the inside lane looking back. In the distance, he could see traffic drawing to a halt. In the foreground lay a trail of wreckage that looked like the aftermath of an open-air rock festival. A pile of rags twenty metres away moved. It was a child.

He fumbled for his phone but was shaking too much and dropped it. Trying to move towards the rags, he collapsed and blacked out.

When he came too, the world was there — people, sirens, radios, and many flashing lights. Someone in green and yellow bent over him; professional, and visibly irked at his consciousness. He managed eye contact, and got a faint nonprofessional smile. "At least, you're OK, I think, though we'll only know for sure when we get you back to the Radcliffe."

Stuart attempted command of his voice but without success.

"Don't try to talk. Just hang in there."

His last view of the scene was of four covered stretchers at the back of another ambulance.

One of them must be Dave, he thought; and the end of dreams.

"You gonna be alright love," said a kindly ambulance driver, as he helped Adele back into her flat. "Can I get you anything? Call anyone?" There's no one to call, she thought, as she tried to smile and say "No, thanks."

"At least, let me make you a cup of tea," he said, setting about the task, as his pager sounded. He ignored it and continued regardless.

"Thank you," she managed, as she forced herself to focus on the simple fact that Austin was alive, though he would remain in hospital for some time, and might never be the same again. But the other fact, that three children, each belonging to some other mother, were dead, overcame her.

Like so many girls in the past, she had found everything technical either boring or simply incomprehensible. She had no idea why her car should behave so very differently, just because it had eight people in it rather than the usual two. All she had done was to look back and try to restore order, after two of them had clambered over the back seat into the boot. When she turned back to the road, she found herself coming up close on another car, so she turned quickly into the outside lane. Her vehicle veered too sharply, but when she tried to correct it, she found it began to veer the other way, even more so. Nothing she did restored control, and the car began to slide and lollop on a path of its own choosing. Touching the brakes just made it slide even more. It left the outside lane, and headed outward.

At first the boys in the back thought this great fun, as though she had done it deliberately, but then they finally did sit still, until the impact burst open the offside door and threw them all into the road.

The ambulance guy made his excuses quietly and left. She had thought to ask him about the other boys but could not pluck up courage. She had seen enough at the site to know it was bad. Austin was not the only one with a head covered in bandages and a restraint. They were still unsure how badly his brain was damaged, and were quite unable to say precisely what the consequences might be. Suddenly, she recalled a radio article about a hospital for brain-damaged children in which every inmate was the victim of a car crash. How often does this happen, she wondered.

Surely, even once is too often.

With no one to comfort her, she broke down and cried.

She didn't move till dusk.

On Monday, the Head at her school phoned and spoke kindly, telling her she would be considered to be on "extended sick-leave" until she felt able to return to work. Adele thanked her, feeling that someone cared. But then the hate and persecution began, with an angry father thumping at the door, shouting "Where's my bloody son." The local press hounded her, by phone and at the door, with questions like why she had loaded seven kids into a car with seat-belts for just four passengers. Then the police arrived.

She answered their questions as best she could, but they just laughed when she asked why, if it was so important, was there never a check on the wearing of seat-belts normally. Though she accepted responsibility, for the life of her, she couldn't see what she had done so wrong. It was new to her that the law now required seat-belts to be worn in the back, as well as in the front, but surely there are many who never bother. She had never heard of anyone being prosecuted. Finally, she was hardly the only lone parent driver to be distracted by children misbehaving in the back. It really didn't seem fair somehow that she was being treated this way. They warned her to expect *prison*. Anyone would think she had used a gun.

It was not as though she had been spared injury herself. The hernia and the broken leg would soon mend, but her mental injuries would take much, much, longer to heal. Depression, anxiety, and panic attacks, did not sound intimidating, but they were robbing her life of all its joy. For the first time, she had contemplated ending it, but immediately realised that, as a parent of a now partially disabled child, she lacked the luxury of that option.

Almost a year later, amidst a hostile media frenzy, Adele found herself in court, charged with causing death by dangerous driving. Despite the kind efforts of witnesses to her character, she was found guilty. The judge said she had created a dangerous situation by failing to stop when distracted. She wondered just how many parents stop when their kids play up, and how many had been prosecuted, aside from the aftermath of an accident.

But it was no use putting up a fight.

The press vilified her. The same irate father turned up repeatedly outside her door to tearfully vent his hatred. Neighbours turned away.

She was banned from driving for six years. Since it was off any bus route, this would leave her unable to get to work. It was too far to walk and there were no safe cycle lanes. There would be time enough in jail to mull that one over. Perhaps the conviction would lose her the job anyway.

She found herself laughing. Things were so bad, they were absurd.

Then she thought of her son. Austin would be "taken into care".

She stopped laughing.

Things were rather better for Stuart, who had made a complete physical recovery. Even the mental scars were healing. Having been raised to "count his blessings", he found himself inundated with them. The experience had awoken him, in every way. Diane had been wonderful, to the effect that he now felt certain of his feelings. That she felt the same way seemed much more than he deserved.

He felt the presence of Dave as strongly as ever. Now that he was gone, what he stood for became a towering influence over his friend, who would never again accept second best, in anything. This would not make him easy to live or work with, but the effort to do so would ultimately be repaid, with interest. Not only did he take up gliding, he went parachuting also, at weekends. To his great surprise and delight, so did Diane.

With help from someone at his college, he set up a trust fund for the surviving boys, in whom he found a mounting interest. The "pile of rags" had survived and bore the name Ashley. But of all the survivors, only Austin had no one. Dave found him languishing in a care home and took him out on Saturdays to play squash and tennis. He even took him gliding now and then. And to visit his mum.

After meeting and talking with Adele, he could find no fault in proportion with the scorn and blame heaped upon her, and told her so. The relief he sensed in her dismayed him, but then he would grow to learn how few in the ordinary world shared Dave's self-confidence. They spoke at length about the underlying cause. He found her astonishingly astute, but then she had had much time for reflection. It mattered crucially to both of them.

You just *had* to understand what had happened, somehow.

They spent little time rubbishing the popular demand for a crash barrier along the central reservation. Now one had been fitted, a head-on collision would usually be prevented, but most trunk roads only had a single carriageway. The majority of such accidents would continue.

They were left with the fact that a minor everyday distraction could cost four lives and a catalogue of horrific injuries, extending to the loss of an eye and three counts of brain damage, including the one lad wearing a seat-belt. There were many such distractions they could think of, such as an attempt to read a map or directions, using a mobile phone, or just a heated discussion or family row.

In the end, they could not "put the world to rights". They were just ordinary folk trying to make the best of the world as they found it. It was for others with greater power to change things.

It was enough that they ended friends.

Two years later he completed his clinical training, married Diane, and took up a post in California.

They named their first child David.

Postscript

Driving as much as we do, we tolerate much distraction, and take it as a 'right' to entertain ourselves in any way we choose. After all, the car is our own private space, and we are unlikely to put up with much interference. While listening to the radio or CD player will not prevent most of us from concentrating on the road and remaining in control of a car, taking one hand off the wheel to hold a mobile phone is now illegal in many countries, including Britain. However, the law is very hard to enforce, and is commonly ignored. In one recent incident, a lorry driver was jailed for five years after running into a car at fifty miles per hour, killing its occupant instantly.

He was sending a text message to his girlfriend at the time.

There is evidence that just holding a conversation can easily become unsafe. Whether on the phone or not, it can significantly distract.

Sources of significant distraction are, in fact, legion.

On making our regular trip to Granny's, my partner once felt moved to keep an eye out for what other drivers were doing, along a busy motorway, at a typical speed of *eighty* miles per hour. Apart from the usual flouting of the law regarding the use of mobile phones, she saw one chap consulting the map held by his wife, and another guy reading a book pressed onto the steering wheel. My personal favourite, and one I had seen before, was the young woman using the rear-view mirror to apply make-up.

We'll leave out some of the more personal distractions. Any curious innocent is directed to Steve Martin's experience in the movie *Parenthood.*

A little less funny for a victim in real life.

3.3 Tailgating

"So, I guess that'll put an end to the drinking, eh. Can't go clubbing when you've a babby on the way," said Gill, of her elder sister, Liz.

"I wouldn't be too sure," said her mum, Jacqui. "I found a bottle of Bacardi round there the other day."

"Doesn't she understand she'll harm the baby? Surely, the doctor's had a word, or the Health Visitor," Gill complained.

"When've you ever known yer sister to take a blind bit o' notice what *anyone* tells 'er?" replied Jacqui, visibly upset. "I've told 'er, I'm 'aving nothing to do with it."

Steve sat on the side lines, more than a bit uncomfortable, but bravely adding "I don't suppose we know who ..."

Jacqui cut him off. "Oh, she was too drunk to remember. In any case, I gather there were many possibilities. Let's just 'ope he wasn't a druggie."

Gill was silent. She'd seen things gradually get worse for years now. First the cigarettes and detentions at school. Then the alcohol and disappearances, sometimes for days. She heard ugly rumours through the school. Enough to frighten her off any of the like. Her sister could defect, but *she* would stay with the system, even if that was none too exciting.

Gill worked hard at school — a "bog standard" compo, with no pretence to scholarly acclaim — and became one of the few in her class to go to university. Her mum had little idea of what that took, given teachers who were not only incapable of keeping order, but who also — it became clear — could not be relied upon even to know their subject.

Despite working all the hours she could get, and counting the pennies, since Gill's dad walked out when she was thirteen, Jacqui had somehow scrabbled enough money to get her a little private tuition. Steve was a research student at Cambridge, who came home often. He made Gill realize the academic limitations of her teachers, and showed her how to study independently. A few hours with him was worth a whole term at school. He would even lend her the books she could not afford.

She learned one more thing from Steve — that learning could offer pleasure and satisfaction, once it becomes proactive. She sometimes wondered whether an institution that called itself a school, but whose principal experience was of being labelled a failure, might have something to do with her sister's decision to opt out. If opting in was so miserable then why not. She still loved, and even respected, Liz. To condemn her would be an unjustified conceit. There was neither hostility nor disdain.

But there was nothing she could do now to save either Liz or their mother. Jacqui was in a position common enough among mothers of girls all down the ages. She faced a stark choice between servitude, from the tender age of fifty onward, and allowing the free-fall of both daughter and grandson. There was no escape, with little help to come from her younger daughter who now had eleven-week-old Lucy to grow and play with two-year-old William. Gill already had her hands full, quite literally.

Despite this, she still heard herself say "Well, perhaps we could ..."

"Don't even think about it," cut in Jacqui. "Don't worry. We'll cope somehow. Maybe it's what Liz needed. Anyway, it's not the baby's fault. He's family and'll get a welcome when he comes, whatever her ladyship does."

"We'll be there," added Steve, keeping eye contact with his partner.

"That's right," she responded, smiling.

"They'll be cousins, and a similar age," he said.

Later, while driving them the twelve miles home, she prompted "I thought men were supposed to hate their mother-in-law."

"One, she's got great legs and a spectacular backside, and two, we're not married, so I'm free to run off with her," he shot back.

"You leave my mother alone! She's too young for you anyway."

He was a little more pensive at breakfast the following morning, but then Sundays never were his forté.

"Truth is," he said, "I'm just not getting anything finished."

She knew well enough when something of magnitude was coming, but could not be too concerned. Lucy still refused to eat almost everything she tried, and William looked like he was going down with something. She had her own share of the problems.

"I was thinking of going and working at SPK for the week. It'd bolster their confidence in me, and I'd get some work done."

"That's OK. I can let the milkman in," she teased.

"We get our milk from the supermarket. I don't think there is a milkman round here anywhere, and the postman is coming up for retirement. I don't think he'd be up for it," came the retort. "Of course, I could go for longer ..." he added.

"No," she said, rather too quickly. "Look, we'll miss you terribly, but I think it's a good idea. You'd get some proper sleep. You can't work without sleep. Anyway, you won't be that far away, and there's the phone."

So, he sorted it out by email — funny how some people even manage a reply within hours on a Sunday — and got her to run him to the station, six miles away, that evening. He'd stay in a spare room in a house shared by two of the younger staff, who could do with some help with the rent.

He was shocked by how much he missed his family even the first night. Children had redefined what 'home' meant to him, more than he'd realized. He chatted with Gill and William, using his mobile for privacy, then read for three hours before nodding off in his sleeping bag.

Though he had worked among colleagues for years previously, it seemed different now that he was used to working alone. He had forgotten how to muster and maintain concentration amidst constant distraction. The environment was one of youth and profound maleness, with a fair amount of mischief. There would be a sudden outburst as someone found their phone superglued down, or a "yikes" as some poor victim sat down on their chair detonating some home-made explosive. He was not spared, and returned from making coffee, one day, to find a hideous alien appearing on his display, every time he moved his mouse.

Lunch-times would be spent wondering the streets of his old university town, revisiting old haunts. He was annoyed to find his favourite bookshop had been unable to compete with the new megastores and had given up its slot to yet another sandwich bar. He resisted being dragged into the daily game of football in the park, but did join in with the Friday pub lunch. Even some of the 'higher-ups' came, who mostly gave up on afternoon production. They knew most of the code got written at funny times of the day or night. They also understood that their "bright young things" were paid for their ideas, and their ability to solve problems and stay up with the pace of change in the internet. Thirty was old in this game. Let them have their fun. Whatever keeps them having fun, keeps them doing what they do.

Derek, the project manager, above all, understood all this, but he also knew the importance of having someone they could look up to — someone not much older, but who had the experience they knew they lacked. Juniors also needed to know the rewards would be there later, and that they would not just get stuck "on the rung" like you did in more secure jobs.

In short, they needed Steve.

There was more to Derek. He kept his faith firmly to himself, but he let it be known that character counted, and that his troop were expected to look out for each other. They would be no "point scoring", and anyone who sought 'guru' status would get short shrift. He liked Steve in particular because he saw stability and values emerging, alongside his technical ability, and he needed those things too. Steve also fitted in; a loner, yes, but not a maverick; a guy who could mingle enough to be part of a team.

"You can come in if you want to, you know," he said, choosing a rare interlude when the youngsters were distracted, just before Steve headed off to the station, to beat the rush. "Really," came the reply, rank with humility, and genuine surprise. Derek knew his own greatest contribution now was bringing in the right people and keeping them. But he was still perfecting the art, and was cautious.

"Cambridge is not such a bad place for families," he risked.

"Just give it a thought, perhaps, and let me know."

"Thanks, Derek. I don't know what else to say. I'll talk it over with Gill, but it's good to be home a lot around now."

"Whatever you need, Steve. We can carry on as we are, or find something halfway. Anyway, mull it over… and enjoy the weekend."

Steve said his farewell and wondered off on foot.

He rang Gill as soon as he was on the train, just to let her know he was on his way.

She met him off the train, with the children.

He mentioned Derek's offer, just to reassure her about the all-important "income stream". She was noncommittal, but made it clear she was happy either way. He had to earn the money. She could look after house and home just as easily there or here, but secretly she wondered why he hadn't taken it up. It was obvious there was much to gain and little to lose. It was boom or bust with IT companies. If yours went bust then you just logged the experience and moved to another one, but if it boomed then the shares would actually be worth something, and you were on your way.

It was a lot less risky than sailing alone.

She would not have been too surprised to learn he had wondered himself. You can't love someone without knowing them pretty well. She knew he knew his weaknesses. He enjoyed the banter, and could retaliate — he was not without wit — but you could never make up for growing up alone, as an only child. This she could never truly understand, though she often tried deleting her sister. It didn't leave much.

To her, his parents had hardly existed. His dad was a shameless workaholic and his mum a quiet reader, who just left him to his own devices all the time. What motivated such people to even bother living together, never mind having a child, she could not imagine. Then they both abandoned him by dying early; she in a road accident, mown down on a pelican crossing by a drunk, and he from a brain tumour, perhaps the result of his excessive travail. Still, they had left a bit of money, that could go nowhere else. It had given them a large deposit on a decent house, which even boasted a small garden. They had also acquired his dad's BMW, free and gratis, when most people they knew had to borrow heavily to finance a smaller car.

All she could do was to try to help him learn *not* to be alone.

But without interfering.

They silently agreed that his working away from home was a bit drastic. Neither of them wanted the normal routine, where he would be a stranger in his own home. Her contribution was to take the children out at least once a day, either shopping or to her mum's, as well as Toddler Group. But Steve would come down and play with them for an hour most days when they got back, in lieu of an hour or two back in his office after they were tucked up in their cots. This was the big reward for working at home.

The week went by as usual, and on Thursday she found herself on the dual carriageway once more heading for her mum's. The three-tonne lorry tailgating her was hardly unusual. As someone who thought through everything she did, she followed several car lengths behind the vehicle in front, which these days was usually a bus or truck. Or at least she tried to. Others would often fill in the gap, forcing her to slow down. Tailgaters would follow her more closely the more space she left or the slower she went. Sometimes, they would attempt to bully her by drawing back and then racing towards her, again and again. Rude gestures were common, even from other women now and then. The other day, a police car had first tailgated and then carved her up, as some twisted form of punishment.

It was all very unpleasant, but not something you questioned. That was just how things were.

Today, she had half given up trying to maintain the gap in front. Then, up ahead, she saw a truck barge inward, carving up another one it had been racing. Its adversary braked over-sharply, causing a chain reaction backward. Perhaps she braked a little sharply herself. There was an almighty bang behind, followed by what seemed an age of unnatural silence. Then their little world imploded amidst a terrifying smashing of glass, shrieking of tortured metal, and screeching of tyres slewing on tarmac.

She awoke, having blacked out, with no sense of the passage of time. Almost immediately she felt physically sick with dread. Then the pain wrenched into her psyche. She could not see for blood in her eyes, nor could she move. Only barely, and with great effort, could she breathe. As the timeless mental weapons of war, evolved over aeons, quelled the pain and cleared her head, she found herself calmly accepting the coming of death. All she could spare her little ones was a silent prayer they had died quickly.

An image came into her head of serenity in her childhood home — a feeling of total security, as her mind built the cocoon that might have saved her from some lesser trauma. But then the creaking silence descended, broken by a groan behind. She sank as awareness grew; they would suffer yet.

The stink of diesel was overwhelming before the faint puff told her the worst. There was fire. She did her best to scream for help but managed only a whimper. Lucy did better, as she emerged from shock. They'll never get here in time, she thought, with a coolness that amazed her.

Firemen, in their 'engine', which would have thrilled William, arrived with surprising speed. Somehow, they had fought their way through the tailback in just fifteen minutes.

"They're alive. I can 'ear 'em," said a bleeding victim from one of six cars now getting caught up in the blaze. "We just need to get those big scissor-things of yours."

"It's not that simple," said the crew leader. "We have to get that fire out first," and he began barking orders and rallying his troops.

"But I can 'ear 'em," the man repeated. Again more quietly, "I can bloody 'ear 'em …" He broke down and collapsed against the wreckage.

He'd go on hearing them, for the rest of his shattered life.

Meanwhile, Steve was making good progress that morning, and was just thinking how odd it was that his family would suddenly pop into his head unexpectedly. He found himself thinking about a game with William at the same time as actually writing code. He had not previously been aware of his ability to give his attention to two things at once. It scared him a bit. It was as though William was there, for a moment.

The phone rang, but it was just Jacqui wondering where Gill had got to. He vaguely heard himself say she might have popped into the mall to pick up some shopping first.

Then came the knock at the door. Hardly anyone ever knocked on the door, but there were occasional traders, some of whom were dubious to say the least. Somehow, it was the wrong knock for that.

He could see the checkered caps through the side-glass, and almost immediately connected it with Jacqui's call. With his heart already in his mouth he opened up.

It was just three-thirty.

All they told him the first day was that all three were dead, and that the crash involved a lorry. He later never ceased to wonder that people could be found to spend their entire professional life dealing with such situations. Only much later would he begin to understand how some actually seek out such work, and that the breed has existed, thankfully, for most, if not all, of history. Of course, these days they can go armed with much wider and deeper knowledge, but still their greatest weapon is time.

The following day, two policemen took him first to the crash site, where they told him just a little more, and then on to see Jacqui. They hugged and cried. He was not yet up to feeling much surprise, not at the presence of Liz, but at her purposefully caring for her mother. It was not the same Liz.

After a while, they felt strong enough to ask about the remains and whether "the authorities" needed formal identification. After all, that's what happens in the movies and TV. PC Jerry Cummings tried a simple "no" but had not expected it to work. Their expressions made it clear it hadn't.

"There was a fire, afterwards," he began, and deftly moved on.

"They'll be kept at the hospital for now. If you like, we'll make arrangements. We'll need to know your church, or whether you want the crematorium. But that can wait for a bit."

Jacqui just stared, along with Liz behind her, but Steve felt a sudden rage burst upward, with no will appearing to quell and manage it. He began to splutter and shake. He leaned forward, pulling his hair, trying to restore control. "I need to see them," he heard himself bellow. "I need to say goodbye. I want to go there now."

Jerry already knew this one was different. Every fatal RTA was hard, he told himself, but he was waking up to the fact that this one was outside even his extensive experience. He turned to his young accomplice and said quietly "I think now would be a good time for tea, love." To Steve, he said "OK," and to Jacqui and Liz he indicated Steve with his eyes. They moved across, and put their hands on his shoulders, either side, doing their best to calm him down. But not for a split second did his eyes drop from Jerry.

The tea arrived, along with some similarly unwanted ginger-nuts.

Slowly, and with practiced ease, Jerry began. "The lorry behind had been following too close. All we know is something made them brake sharply, and it went into the back of them. What happens then is that the heavier truck flies up in the air, as the car slows down. Often it rolls to the side, but this time it kept going straight ahead."

He paused, using the tea as a device. It could have been Jackanory, with added technical insight. Jerry would use anything to distract from the reality. It was as necessary for him as it was for his audience. Strangely, and to his great pleasure, he found he had become adept at telling stories to his grandchildren. There was the same need for patient decoration.

"I'm afraid this time the offending vehicle landed on top of your wife's car." How he would have loved to add "there's no way they could have survived," but he knew better. Instead, he continued "They were crushed. Later, there was a fire, so I regret to say there are no bodies to identify. It's enough that we can identify the car and be sure that no one else would have been using it. We just came to the address of the registered keeper."

A gentle deceit.

Steve was staring at the floor. Soon, he would wake up and shudder, after one of those dreams that seem real. Surely.

After the second night sharing the house with a minder, he began to realize there were significant practical problems that could not be ignored. He was also wise enough to know, objectively, that he had to take care of himself. He could not bear to be alone in the house at night, nor could he continue availing himself of the generosity, above the call of duty, of Jerry, who had a wife to go home to. Worse, he didn't think he could stay in the house at all, yet there was much in it that was priceless to him.

The phone rang. Out of habit, he answered it. Derek could not have known how inappropriate was his innocent joviality. Steve told him the news, apologizing for his inability to work for a while. Stammering out a summary of his progress, Derek interrupted him with a simple "Stay there. I'm coming." Two hours later, just before lunch, a car pulled up outside, spilling Derek, and the two guys from the house he had stayed with, in Cambridge. They said little besides "Get your things. You can't stay here." He hesitated long enough only to shake hands with and thank Jerry, and to call Jacqui to let her know where he would be. Liz answered, and offered to empty the fridge and freezer for him.

He took little more than a few clothes, his phone, and his laptop.

Over the next few months, he came to be very close to James and Aman, his two housemates, who always made sure one of them was at home to keep an eye on him. Jerry had surreptitiously passed Derek a card with the number of a charity, specializing in the care of victims of psychological trauma, but it was not needed. As luck would have it, Derek was a Samaritan and had both professional training and some experience, though of nothing quite like this. He took time out, both to brief James and Aman on what to expect, and to contact a specialist at a London hospital for advice. They steeled themselves for the occasional bouts of irritability, rage, and hypersensitivity. They watched for the signs of impending self-harm, or exposure to unacceptable danger. But these never came. As veterans of countless savage wars know so well, you can never tell who can take the punishment until it comes. More fall after the battle than do so in it. You just can't say who will stand until much later.

In years to come, he was to wonder what might have become of him, had not a Samaritan passed by. A couple of generations ago, it would have been family who would have picked him up and taken care of him. Failing that, it would have been neighbours. But he didn't have any family left, and his neighbours barely knew his name. After neighbours, it would have been the Church, led by the local vicar. But he had no faith and there was no church anywhere near. They didn't build one when they built the estate. Vicars were the stuff of old black and white movies from the 1950s.

At least Jacqui had Liz, who came good when needed. To his shame, he would never have guessed.

After six weeks, Jerry called to say that the lorry driver had been charged with causing death by dangerous driving and gave him the date of the court appearance. Time to learn a little more. Knowing so little of what his family had suffered was causing him increasing trouble. It allowed his imagination too much liberty to fill in the gaps, whether awake or asleep.

"Any drugs, drink, ... past convictions," he asked.

"I can tell you his blood test was clear."

After a second he added "This was just plain bad driving. That's all. Following too close. Taking no account of what can happen. The specialists talk of funny effects you can get down a chain of vehicles when one makes a sudden change in speed. They say the change can get 'amplified'. You're technical. You'd probably make more of it."

"Can you get me the report?"

"No, sorry. Police Accident Reports are part of criminal investigations. They're even exempt from FoI claims."

"FoI?"

"Freedom of Information Act."

"You mean there's no public record of what happened?"

"This is England, my friend. Everything's always been a secret. Still is, I'm afraid."

Jerry's guard was down. He felt more that he was speaking to a friend than playing mentor now. "You could try the Coroner's Court," he said, instantly realizing his mistake. But then, he thought, perhaps he has to know the truth sometime. Perhaps he's ready.

Immediately afterwards, he phoned Derek.

The very next day, Steve set out for the train, refusing Aman's offer of company. "I could do with a little time now on my own," he said. "I'll be OK."

He had called to check the record was available. The inquest had been and gone, but they would say nothing on the phone. Only when he put the phone down did he wake up to what had been nagging him from the back of his distraught mind from the beginning.

When he arrived, he could not help thinking how little we are all aware of these institutions, now that births, deaths, and even marriages, are so much less common. His professional mind could not help wondering how much progress had been made digitizing all this stuff. He then wondered how much that would further diminish people's lives, condensing love and loss into mere numbers. The cold technical definition of 'information' spat into his mind, from his college days — "that which resolves uncertainty".

It could not better sum up what he sought.

He seemed to be getting preferential treatment and was allowed privacy in a small meeting room, where they delivered the document. The young woman who brought it seemed at first officious, but then he noticed her nervousness. The seconds passed as he found his way through, until he at last landed on the medical report. His worst subliminal fears were confirmed. "My God," he heard himself mutter. "No."

He broke down violently.

A hand appeared on his shoulder. He looked up, startled, to see Derek.

"You knew, didn't you," he said.

Derek braced himself for hostility, that did not come.

"Yes, I knew," he replied.

"They burned to death," he sobbed quietly. "It says 'all three died from severe burns'."

"That was the cause of death. They may still have been unconscious from smoke and loss of blood," said Derek, doing his best. "Whatever happened, it can't be changed now. It's done," he added. "Life is for the living."

After a while, he took them to a café and bought coffee. He drank half his while Steve sat in silence.

A month later, Jacqui phoned. "It's a boy," she said. "Liz has had a boy."

"That's wonderful," he heard himself say. He was surprised by a growing sense of warmth.

"Guess what she's going to call him."

"I have no idea."

"Tyrone! What do you think?"

He actually heard himself chuckle. "Love it. It's Irish isn't it?"

"Oh, that's in the blood anyway, I'm afraid. Anyway, she wants to have him christened, and you to be a godfather."

"Has she got religion?" he replied, dumbfounded.

"Nah. Just come over all traditional. Next thing she'll go and get a sensible fella."

"Well, I'll tell the boys here to watch out," he said, realizing that he was coming close to humour. But hey, what of it. This *was* good news.

There could still be joy.

He became a true uncle, visiting regularly, and buying the biggest presents at birthdays and Christmas. He even started a secret savings scheme for the mite's education and future. After all, he was the only one who could. After much thought, when the lad was two, he bequeathed Tyrone all William's toys.

At Derek's suggestion, he had decided to sell the house.

Liz had not just cleared the fridge and freezer. She had carefully packed all that was precious into boxes and stowed them in Jacqui's loft, a little more secure from theft.

He went back just the once, on his own, to say goodbye. Left to himself, he might have kept it exactly as it was that day, as a shrine to all that they had had. But it would be a morgue, and a vast waste of assets. He sold it, and every last thing in it, and put half the money in Tyrone's fund, along with the return on the insurance claim against the loss of the car. The rest he invested in a small flat in Cambridge, into which he moved along with Aman, after James took a new job in London.

His work carried him, more than anything else, along with many distractions. He played squash, and began running marathons for charity. Having chosen his flat carefully, he walked to work, and to shops, cinema, and restaurants. Sports centre and rail station were just a short bike-ride away. He never drove a car again.

It would be a long time before he'd trust himself alone for any length of time, and he stayed clear of women. Aman being gay was useful in that sense. They both got a laugh out of the inevitable misunderstanding and innuendo. Steve was decidedly 'straight', and was far from Aman's type.

Derek's charity was well rewarded. Steve brought not only leadership and credibility to the company, but business as well, from contacts gained while contracting. He became both manager and shareholder — a combination that went down well. As a result, he could afford new distractions, some of which worried his friends a little. Derek introduced him to sailing, in the Mediterranean — a duck pond, he called it. The following year, Steve took time off to crew a mere thirty-foot boat across the Pacific. They survived the obligatory cyclone to arrive safely, though with some damage, in Tahiti.

Skydiving became a passion.

It was simple enough. Gill, William, and Lucy, had looked death in the face. He would join them.

Postscript

Tailgating — the practice of following dangerously close to the vehicle in front — has become pandemic in England. While it is recognized as a form of "dangerous driving", which is specifically illegal, it no longer attracts any attention by the police; at least, not until after a fatal accident has occurred. Along with abuse of speed limits — also pandemic — its prevalence denies any claim that the law on the road is effectively enforced.

While heavy vehicles are designed to share stopping distance with lighter ones, they will do much more damage in a collision with a car. The innocent in front will die, while the lorry driver survives.

But not always. One accident, that occurred recently almost within earshot of the desk at which I write, involved five lorries, which piled into the back of each other. The driver of the *fifth* lorry died, while the others survived, though two were seriously injured. (The victim had lost his fifteen-year-old female cousin the previous month in another road accident.)

Accidents caused by lorries tailgating are no rare event. Just six months before, within half a mile on the opposite carriageway, the very same accident occurred for the same reason, involving just three lorries. As luck would have it, no one died that time.

Tailgating is not only directly responsible for many accidents, it also makes *any* accident much more lethal. It can turn a minor shunt into a major pile up, multiplying the casualty figure. Quite apart from this, it is inherently criminal — essentially constituting a threat with a lethal weapon.

Perhaps worst of all, as I attempted to illustrate in the story above, tailgating is self-propagating. If left unchecked, it becomes a behaviour that is eventually forced on all of us, whether we like it or not.

Like any other form of bullying.

Post-postscript

After identifying the accidents I wished to portray, and learning all too much about them and their consequences, I did my best to avoid any further immersion. Sadly, I was still confronted with a repetition of this one, almost to the last detail. The same crime claimed the same innocent victims.

4

Sustenance

4.1 Energy: There's a lot of it about but it's hard to carry

Revolution

No one prior to the Renaissance, and few during it, could have guessed that by lighting a fire under something you could make it move, though the ancient Greeks came close. After all, Democritus had postulated that matter was made of atoms, and Heraclides that planets orbit the Sun. Three centuries before Christ, Eratosthenes not only knew the Earth was round, he measured its circumference (correctly) with the aid of nothing more than a stick. (Consider how he did this an exercise in lateral thinking.)

Greek triumphs were not limited to leaps of science. As with us, their technology progressed hand-in-hand. Among a pile of booty, shipwrecked off the island of Antikythera, has been found the remains of a mechanical calculator, the like of which was not to be seen again until the 17th Century. Writing sometime in the first century A.D. at the world's first true university, Hero of Alexandria described a cauldron passing steam to a hollow ball mounted on a pivot, causing it to rotate in the fashion of a rocket. He thus demonstrated both an understanding of the power of steam and the essential idea of turning fire into motion, at one and the same time.

Hero was only reporting the work of others. Academic credit seems not to have been fought over quite as much at the time, or perhaps was just less well defended. Strato, whose name rose to great heights, is now considered responsible for many of the pneumatic devices Hero described, together with the possible origin of true scientific experiment. (It was Strato who correctly asserted the primacy of *observation* over logical argument — the very essence of science.)

All the components of a useful working steam engine can be found among the devices reported by Hero — the boiler, the piston and cylinder, and the crank to convert linear to rotary motion. Sadly, the Romans arrived before anyone brought the pieces together. We can only wonder how history would have read otherwise.

Throughout the Renaissance, the pieces remained apart, despite the demand upon technology in the pursuit of war. Circa 1500, Leonardo da Vinci experimented with steam power, according to an illustration collected in the *Codex Leicester*.[1] He also drew a "steam blower" and even made a steam-powered canon. Leonardo clearly understood the power of the terrific expansion that takes place when water turns to gas, but he failed to associate it with propulsion, despite drawing a credible automobile.

Necessity is the mother of invention. If you were a miner, you needed water pumping and your mineral hauled out and away. From the West Country of England, Thomas Newcomen brought the first steam beam engine into use for pumping water out of mines in 1712. This brought two components together — the boiler plus the cylinder and piston.

The idea of mounting such a machine on a moving vehicle would have been the cause of only mirth. Once the final component — the crank — was added to permit rotary motion, it made rather more sense to keep the engine stationary and simply pull a wagon along. It was clearly daft to waste energy carrying along a machine that weighed more than your load. It was downright preposterous to lug the coal along too. Not only is coal rather heavy, but early steam engines were only able to use about 1% of the energy it held. In other words, you would have to carry rather a lot of it.

Pretty soon, however, the power of steam was needed a little too often for the engine to remain stuck in one place. Curiously, it was another soul from the West Country that was to complete the picture, though this time from across the border in Cornwall. In 1796, Richard Trevithick produced the world's first locomotive as the Industrial Revolution gathered pace.

Trevithick was to prove a truly great innovator.

---

1 The *Codex Leicester* was purchased by Bill Gates in 1995.

His *Puffing Devil* (1801), on which he took friends for rides, used a spinning flywheel to smooth its operation. The *Penydarren* (1804) ejected its steam up the chimney and so drew more hot gases through the boiler and more oxygen into the fire. It proved capable of hauling ten tons of iron and seventy passengers from the Penydarren ironworks in Merthyr Tydfil, Wales, nine miles to a canal.

Sadly, its success was limited. After just three journeys, it was retired. At seven tons, it was so heavy it repeatedly broke its cast iron rails.

But the world was changed forever.

Locomotion was borne, and soon spread far and wide.

How hard it is for us to understand just how wondrous it must have been, to those used only to the power of muscle, to be carried along by a machine, mile after mile, at speeds no horse could match, let alone sustain.

Railways were originally intended to carry raw material for industry — the very first line carried cotton from Liverpool docks to mills in Manchester — but they were also to enable far wider distribution of manufactured goods. The effect of rapid, reliable, high-capacity, transport on industry was to stimulate a massive growth in production, enabling not just greater and faster supply but also far wider distribution.

In the first half of the 19th Century, many would ride the railway (if they could afford the fare) simply for the novelty and exhilaration. Once people discovered travel they took it up with alacrity. Travel itself became a product, and one in great demand.

None of this would be possible without a supply of energy.

The first trains to cross the continent of America burned wood but the source soon became too distant. After a fair bit of deforestation, coal became the staple, and remained so, all around the world, for over a century.

Trains could haul coal just like any other commodity. The effect of more industry was yet more industry and more transport. The effect of more transport was yet more transport and more industry. Increasing supply caused increasing demand with the result that growth of both production and consumption became exponential.

Exponential growth has yet to be reconciled with any finite resource.

Some necessary materials were not available close by. New steamships could scour the globe for them. Because Britain had these first she would establish an enormous empire — the greatest and richest in the history of the world — guarded by a correspondingly large and expensive navy.

Many a war would be fought over resources. Even former friends would become enemies, when growth was threatened by supply. Hitler may even have inadvertently prevented a transatlantic war. In the 1930s, America engaged in a naval "arms race" and was ready to blockade British Malaya, as it was then known, over access to rubber.

Britain's empire guaranteed her trade as well as supply. Industrial production was only economic with the largest possible market. The Royal Navy "ruled the waves" and effected control over global trade. Pacts and treaties were made, some public and some secret, but all would favour the empire "on which the sun would never set".

War was perhaps inevitable.

History will not record the first two world wars as separated in any significant way. At the end of the first, a French general predicted the onset of the second. He was only six months awry. The Great War was no ordinary war. In effect, it persisted for more than three decades, cost the lives of tens of millions, and brought sunset to the British Empire.

Energy was *not* the cause of the conflict that arose from the Industrial Revolution. This was merely a matter of luck. Coal is well distributed around the world. Most industrializing nations had a plentiful supply at home. (An exception was Japan.)

The same is true for the new industrial nations of China and India. They have enough coal to keep their industry going for centuries. The problem is that coal is not a 'clean' fuel, nor will it ever be cheaply extracted. Burning it liberates toxic fumes which cause respiratory disease that disables and kills by the thousand. Its sulphur content produces "acid rain", which destroys crop and forest. And then there is the carbon dioxide created.

Before we consider each possible alternative, we first reflect on the origin of energy and where we might look for it, given the extent of our current knowledge. As we shall see, there's a lot of it about.

Manna from heaven

Perhaps the most familiar unit of energy is the 'kiloWatt-hour' (kWh) that appears on our electricity bill. A 1kW heater is usually enough to keep us warm in a typical room on an average day. We measure the effectiveness of a heater by its power, and the cost of using it in terms of the energy it gets through. The two are of course intimately related. Power is just the rate at which energy is consumed (energy per unit time). Leave the heater on for an hour and you add a unit (kWh) to your bill.

An electric heater serves as an example of the *conversion* of one form of energy (electricity) into another (heat). Your local electrical power station is yet another. One thing we're pretty sure of now is that there is exactly the same amount of energy in our universe now as at the time of its creation. Energy is never either created or destroyed. All we can ever do is convert it from one form to another, according to our needs. In so doing, we spread energy around, or *dissipate* it, typically in the form of heat. Nature is busy doing the same, via stellar radiation, for example.

When we 'consume' energy, we merely downgrade it.

The process is not reversible. At some point in the far distant future, all fuel of every kind will have been consumed and the universe will die, but there is no cause for alarm. "Heat death" is a *very* long way off.

A typical power station converts the chemical energy in, say, natural gas into electricity with an output of perhaps a gigaWatt (gW, a million kW). Current global human power consumption, of all kinds, amounts to about 12.5 teraWatts (tW, a thousand gW). The output of the Sun is around 385 *thousand billion* tW. Stars maintain this for billions of years.

The Earth is irradiated with about 175 thousand tW.

We might respect nature and be grateful.

In 1820, Hans Christian Oersted and Anne Marie Ampère discovered that an electric current produces a magnetic field and can thus cause motion. Eleven years later, Michael Faraday demonstrated the reverse — that motion through a magnetic field could induce an electric current.

Physics seeks to explain multiple phenomena with a single principle. By 1865, James Clerk Maxwell had shown that the electric and magnetic fields are manifestations of a single essence — the electromagnetic field. The principle behind both the electric generator and electric motor was established.

One consequence of his theory, among very many, is the inevitability that electromagnetic (EM) waves would radiate from any disturbance. It quickly became apparent that light and radiated heat were merely EM waves which differ merely in their length (or frequency).[2] Two phenomena, perceived quite differently, could now be understood in the same way.

This is a classic example of physics at work. Physicists seek to reduce the number of variables needed to explain all we observe. We do not yet know the ultimate end to such reduction — whether all that we see can be explained in terms of a single essence, or whether many ingredients will remain. Whatever its terminus, physics can never explain *why* we are here at all. A universe of utterly timeless, silent, cold, dark, night would break no 'law' of science. Neither would the complete absence of any universe. Such laws merely express what cannot (yet) be explained through reduction.

Physics only marks the tapestry of Creation. We can claim no explanation for the *particular* universe in which we find ourselves. Maybe Darwin was more right than he knew, and natural selection applies even here. Perhaps we simply inhabit the only world fit to survive — that 'works' in every respect, and thus is here to witness, and bear its witnesses.

Whatever its origin, it is one that bequeaths a rich energy legacy.

Radiation can be regarded as just another form of energy, transmissible through empty space, allowing the Sun to warm the Earth. To make EM radiation, something with electrical charge has to wobble. To be precise, it must accelerate.

The American physicist Richard Feynman was once asked what shred of knowledge we should most strive to preserve should civilization fall. He replied "that everything is made of atoms". A star is no exception. While atoms do not have any net charge, their constituents do. Electrons possess thousands of times less mass than the atomic nucleus. As a result, they may be expected to wobble a lot more — enough to account for almost all of the radiation we see.

---

2 Wavelength (distance between crests or troughs) and frequency (number of complete waves per second) are inversely related — as the one increases, so the other is diminished.

When the atoms in a gas wobble about, we perceive it as 'hotness'. The more they wobble, the higher the *temperature* of the gas. As a gas gets hotter, its atoms eventually break apart, leaving 'ions' and electrons. Measuring the temperature of the Sun, or at least, its outer layer — the 'photosphere' — is not as hard as you might expect. Since a human first lit a forge, we have known that as something gets hot it begins first to radiate heat and then to glow, with a colour that changes with temperature. The process begins with a dull red and ends with white. Colour emerges as merely our perception of the wavelength (or frequency) of EM radiation.

When we bathe in the sunshine of a summer's day, we are enjoying energy that has crossed space from the solar photosphere, where it was generated by raw heat, generated in turn by something else.

According to theory, our universe began with an almighty bang at a single point in time. As in any explosion, matter is hurled outward and gradually cools down. After a while, particles condense to form atoms, mostly of hydrogen, where a single electron orbits a single proton. This too then condenses into a gas, two atoms to a molecule, that begins swirling around.

Later, as the maelstrom cools, denser patches emerge, and gravity begins to attract more material into each one, which then might just collapse into a blob. If the blob is big enough, gravity will squeeze the material in the centre so tightly that molecules are battered back into atoms, and atoms back into particles (protons and electrons).

Sometimes when two protons smash into each other, one disintegrates into a neutron, of about the same mass, and a positron that carries away the electric charge. Protons and neutrons may then *fuse* to form 'deuterium', and eventually helium (two protons and two neutrons) liberating enormous amounts of energy, in the form of heat and radiation.

An aptly named "strong nuclear force" is able to bind them together.

If you throw a proton at another proton, they will not collide. Anyone who played with magnets as a child knows that like poles repel and opposites attract. Static electricity works the same way. Protons all share the same positive charge. They repel one another. However, if you throw hard enough, you will find they do, in fact, stick together. Luckily for us, there is another force at work in nature, invisible on the ordinary 'macroscopic' scale.

Electric repulsion acts over a much greater distance but is very much weaker. Two protons flying fast enough towards each other may overcome it. For each one, it's much like racing up a hill, only to fall into a deep well at the top. Many protons may fall into the same trap. In so doing, some get smashed into neutron plus positron. New 'elements' are formed by collisions between nuclei of increasing size. Eventually, nuclei comprise so many protons that nothing else with charge can get close enough to bind — not even a single proton. Only uncharged neutrons can add to their mass.

Iron is just the ash of nuclear fusion. If you fuse two iron nuclei, little energy is liberated. Making elements heavier than iron requires a "neutron-rich" furnace; for example, the core of a collapsing star, just before, or just after, it explodes. This is why they are comparatively rare.

A relatively minor explosion is called a 'nova' and leads to some very beautiful objects at which to peer through a telescope. A nova can destroy all life not just in its own planetary system but also in its neighbours as well, though they are so far away it takes even light years to get there.

But even this falls far short of nature *in extremis*.

A 'supernova' can outshine an entire galaxy of perhaps a hundred billion stars. Such a calamity does not, thankfully, happen often — perhaps just once every thirty years per galaxy. Novae and supernovae are responsible for spreading around everything heavier than helium. No novæ, no planets; at least, not the hard ones you can walk on — the ones on which you can mine metals and develop technology.

We are indeed stardust. So is our car.

Heavy nuclei effectively store the energy it took to forge them. We have seen how heat can be converted into EM radiation, and back again. It can also be converted into "binding energy", at either the molecular or nuclear level. As was first so tragically demonstrated at Hiroshima, the energy stored in the nucleus is vastly greater than that in any molecule.

The legacy of a dying star is both metal and fuel.

When an object truly falls down a well, it acquires energy. A proton flying fast enough to overcome the 'hill' of electrostatic repulsion does likewise. It becomes strongly attracted to its target, just as the object falling down a well is attracted to the centre of the Earth.

From hydrogen up to iron, fusion liberates energy.

As particles accumulate in the target, the height of the hill and depth of the well vary. The former grows with the number of protons in the nucleus, and the latter diminishes with the total number of nucleons (protons and neutrons) — as though the well were filling up, leaving less far to fall. The base of the well can eventually lie *above* the foot of the hill. Beyond this point, energy is lost to the reaction, and stored in the nucleus.

Fission can later release this energy.

A very large nucleus is unstable and easily shattered.

Imagine a number of solid rubber balls crushed together extremely tightly with something that acts like shrink-wrap but with the strength of steel. Each pair of balls are similarly wrapped together. Now suppose they are struck with another ball travelling fast enough to burst the wrapping. The contents will fly apart with tremendous violence. The smaller the debris, the faster it will fly. The smallest will possess enough energy to burst a similar bundle, and so on, forming a chain reaction we commonly call *fire*.

Nuclear fire burns much hotter than the molecular (chemical) kind just because the nuclear bond is far stronger than that between atoms.

Fuel is simply any material that either stores binding energy or (in the case of fusion) potential binding energy. The latter kind will clearly possess less mass and thus be easier to carry. For this reason, fusion (chemical or nuclear) is much preferred over fission.[3] Sadly, the former (of either kind) is much harder, and more expensive, to ignite and maintain.

There is no chain reaction that can sustain fusion.

Like latent stars, rocky planets too get extremely hot as they condense. Over millions of years they cool down. Convection or conduction brings the internal heat to the surface where it radiates into space. Yet there is still evidently molten rock beneath us.

It has become clear that continents move. Their movement is fast enough to be measured using modern instrumentation. In fact, continents and oceans ride on 'plates' into which the surface crust of our planet is fragmented. These collide, forming mountain ranges, and rub against each other. Friction between them releases colossal amounts of heat.

---

3 Chemical fusion is achieved between hydrogen and oxygen in a "fuel cell", generating electricity and producing only water as 'ash'. We will have more to say on this later.

The energy to drive this movement, and to liquefy rock, must come from somewhere. It seems very unlikely to be primordial heat from gravitational accretion. Knowledge of radioactive decay allows us to reliably tell the age of the solar system. Four and a half billion years is long enough to cool just about anything. Tidal forces are a possibility, but the Moon is not massive enough, and nothing else is sufficiently close, to provide anywhere near the quantity of energy we see. Nuclear fission is now the accepted explanation, but the truth is that we simply do not know the source.

It's kind of hard to go and look.

Naturally occurring nuclear fission reactors have been found, but the quantity of reactant, beyond criticality, would have to be enormous. The only possible evidence is helium, found emerging from deep below the surface. Two distinct isotopes are found; one with a single neutron, the other with two. Their ratio differs with that found in the atmosphere and is consistent with subterranean fission of thorium and uranium. The ratio observed suggested to at least one investigator that the reactor might be close to shutting down. This would bring down the Earth's magnetic field which is our primary defence against lethal solar radiation.

Whatever its origin, geothermal energy is plentiful, to say the least.

Were you to hold a single electron and a single proton motionless, in the vicinity of each other, and then release them, they would fall together and there would be a bit of a bang. Theory and experiment agree conclusively that particle and 'antiparticle' annihilate each other with a truly staggering release of energy, dictated by Einstein's famous $E = m.c^2$.

A matter-antimatter reactor remains plausible, but entirely beyond our technology. We have yet to master chemical and nuclear fusion.

Natural electron-positron annihilation is most unlikely.

Electrons are light and easily propelled. They are very likely to be in motion; in other words, they will possess energy. Like a comet falling towards the Sun, an electron will keep its initial energy and fall into an orbit around a nucleus; the greater its energy, the more removed the orbit. A better image of an electron in an atom is that of a diffuse three-dimensional 'shell'.

If this picture told the whole story, two hydrogen atoms would simply repel one another because they appear to each other as a ball of like charge. At a distance, this is indeed the case. Up close, however, the electron in one atom can still perceive the proton in another. Surprisingly, there are stable configurations in which atoms are bound together.

We would not be here otherwise.

Imagine the electron shell as a liquid, complete with waves. Any wave would have to be continuous around the surface. Such rules do indeed seem to apply, and constrain the 'shape' of any such shell. Each configuration can accommodate a certain fixed number of electrons. It is then said to be full. The number of vacancies in the outermost (lowest energy) configuration of an atom is called the 'valency' and dictates the number of other atoms to which it can bind. Every element thus has a characteristic valency.

Hydrogen is 'monovalent' and carbon 'quadravalent'. Hence one obvious possible chemical compound is carbon tetrahydride — written CH_4, and better known as 'methane', or "natural gas". To 'cook' methane, you simply introduce carbon to hydrogen and warm up the mix. Heat implies that atoms rush about, whereupon they bump into each other. When the material is hot enough, atoms overcome their electrostatic repulsion and bind.

There are many stable hydrocarbons, and quite a few that are far from stable. Unfortunately, it is almost as easy to cook *explosive* as it is fuel.

The reason hydrogen is such a wonderful fuel is that its very first electron configuration, available for sharing and thus bonding, is also the one that stores the greatest energy. Any element with more nuclear protons will already have that shell full. Hydrogen can thus store more energy per atom than any other element. Being also the lightest element, gives it by far the highest "energy density" of any fuel.

Fusion of hydrogen with oxygen propels the largest rockets.

Carbon makes a good vehicle for hydrogen bonds because it can accommodate four at a time, yet is still quite light itself.[4] Nature has had plenty of time to explore and find the best options for fuel. It has chosen hydrocarbon, and manufactures it from a sustainable supply of both raw material and energy — water, soil, and sunlight. We simply exploit that which it made earlier and has not yet got around to recycling.

4 A carbon atom comprises six protons, six electrons, and (typically) six neutrons. It is thus said to have "atomic number" six, and "mass number" twelve.

Hydrocarbons are also easy to ignite. Break one methane molecule and you get missiles in the form of hydrogen atoms. Being light, these fly off fast. There are also four of them — ideal for starting a chain reaction.

Hydrogen, however, ignites *ten times* more easily.

We have been given the perfect chemical fuel, in hydrogen. The problem is that it has all been burned already, to make our oceans.

If a day were spent as witness to the evolution of our world thus far then life would arrive only around one minute before midnight. But life can utterly transform a planet. For a start, more solar energy is absorbed, instead of being reflected back into space. About one percent fuels the synthesis of living material from the carbon in the air, above both land and sea. Less carbon in the air means less energy retained by the atmosphere. The temperature falls and the weather calms. Oxygen is released allowing combustion.

Living things die. They usually then rot or burn and their carbon returns to the air, to feed new life. Any such material (wood, for example) is termed 'biomass'. Sometimes dead things become buried, ever deeper, until they meet the magma below and get ejected back into the air volcanically. The route via immediate decay is called the "short carbon cycle"; that via the volcano, the "long carbon cycle".

Occasionally, the remains from life that get buried underground enjoy a highly unusual history. First it gets heavily compressed, then it is cooked by subterranean heat. If the circumstances are just right, deposits of coal, oil, or natural gas, are formed — collectively known as "fossil fuel". After the cooking, and following further distillation, fossil fuel stores much more energy per kilogram than biomass.

Chopping down a tree for firewood makes no difference to the planet. Whether it falls and rots or is cut down and burned, its carbon returns from whence it came. The short cycle will turn complete.

Digging up and burning fossil fuel is altogether different.

In summary, there is but one ultimate source of energy not already dissipated into the void[5] — the primordial one locked up in every atom around us, that

[5] This is not quite true; there is also gravitational energy. For example, planetary orbits sustain energy which can be tapped by passing spacecraft. A 'slingshot' manoeuvre can result in a much faster spacecraft and an infinitesimally slower planet.

can be released by nuclear fusion or fission. All non-nuclear options on Earth are derived from solar EM radiation. These then reduce to solar-direct, biomass, and fossil fuel. The difference is really only one of timescale.

While geothermal energy may indeed be the result of nuclear fission, it can be regarded as an extremely large and independent source of supply. Easy access is restricted to relatively few regions around the globe, though it is possible that deep (10km+) drilling technology may widen access significantly in the near future.

'Solar-direct' refers to any method of recovering useful solar energy in a manner more immediate than biomass. There are many such methods, all of which constitute a *sustainable* (or 'renewable') energy supply. These are of such importance that an entire subsection below is devoted to them.

Finding a plentiful and sustainable supply of energy is one problem. Delivering it safely and efficiently to the point of consumption is another.

There is a lot of energy about, but it can indeed be hard to carry.

As our ancestors knew too well.

Let it roll

To propel an autonomous vehicle (*i.e.* an automobile), we need to carry fuel along with it. Direct connection to a power supply is not an option. We therefore seek a fuel that carries as much energy per kilogram as possible. It's bad enough having to carry an engine around, without adding an excessively heavy fuel tank.

As we have seen, simple hydrogen is the chemical fuel with highest energy density, because of the strength of the bond between the two atoms which make up each molecule, and their small mass. A poor second best is hydrocarbon, where a much weaker bond is associated with at least eight times the mass. Third place, far behind, is occupied by molecules held together by relatively low-energy carbon-carbon bonds.

Unfortunately, availability follows the reverse order.

Hydrogen does not occur naturally in any quantity on Earth. If it did, it would soon disappear. Being lighter than air, it would float up through the atmosphere where it would either oxidize to form water or eventually leak out into space. It is therefore not immediately available to consume but must be synthesized as a *secondary* fuel. The most obvious source is water, given that we happen to find ourselves on a very wet planet.

Electricity can bring about the fission of water — a process called 'electrolysis'. Chemical fusion of hydrogen and oxygen will liberate it again. This is literally perfect for powering an automobile. First, the only vehicle emission is water vapour, which nobody minds. Second, electricity may be efficiently converted into traction. An electric motor can be more than 95% efficient. Third, hydrogen is the lightest of all elements. In principle, little energy need be wasted in carrying the fuel with you.

Hydrogen and oxygen can be reunited in the "fuel cell" mentioned earlier, which liberates not only water, but electricity as well. A fuel cell manages to convert 45–60% of the chemical energy in hydrogen into electricity. Both cell and motor are also *quiet*, unlike the internal combustion engine (ICE) which is only about 25% efficient at best.[6]

Any vehicle must minimize its "mass ratio" — the ratio between the mass of the laden vehicle and that of the load alone. It's no use designing a hundred-kilo car that weighs a ton once you've filled the tank.

Energy density has always ultimately governed the choice of fuel. That of hydrogen is nearly three times that of petrol. That of petrol is about half as much again as that of coal. Oil is less of an improvement on coal, but the convenience of a liquid meant change was inevitable. Operating costs shrank dramatically when men and shovels gave way to pipe and pump, all the way from extraction to combustion.

While the energy density *by mass* favours hydrogen above all other fuels, it is a different story when measured by volume. Hydrogen is the least dense of all gases. Just 3kgs occupies 36,000 litres at atmospheric pressure.

Much effort is going into solving this problem.

There are two dominant approaches. First, other liquid fuels, such as ethanol, that can be refined from fossil fuel or biomass, contain much hydrogen which can be recovered fairly easily, though with varying ef-

---

6 The efficiency of any such engine is limited by the Second Law of Thermodynamics, according to the ratio of exhaust to combustion temperature. Burning gasoline leaves an upper bound of 68%. Actual combustion temperature is constrained by the cylinder block material. The use of steel for cylinder and block, together with the need for a gearbox and transmission, reduce this limit to around 37%. However, even a car capable of 14kms/l (40mpg) at 100kph (63mph) is less than 20% efficient, neglecting its mass ratio.

ficiency. Second, hydrogen can be adsorbed into metals to form hydrides, yielding about half the volumetric energy density of petrol.

There may be a better, though rather exotic, solution. Nanoscopic tubes constructed in geodesic style from carbon atoms have been claimed capable of storing hydrogen with a volumetric advantage as great as the one by mass. Carbon 'nanotubes' appear able to store *three and a half times* as much energy per litre as gasoline (~32kWh/l compared with ~9).

Exotic tends to mean expensive. Fuel cells and carbon nanotubes are each currently both. However, the same could once be said of the ICE. Mass production on a similar scale could perhaps one day repeat the same success. But many gigantic obstacles remain, not least finding the hydrogen. Manufacture by electrolysis would require doubling electricity production. Replacing the existing gasoline distribution network could cost up to $15bn in the US alone. Raising the investment requires prior demand — a classic "chicken and egg" problem.

Two more problems remain before a market can rise. First, fuel cell power output does not yet approach that available from a cheap ICE. Second, range, with a tank of reasonable size, remains wholly uncompetitive.

One possible selling point lies in the fact that the power output required for a car (50–100kW) exceeds that required for a typical house (10–30kW). When not travelling, a fuel cell vehicle could power the home.

Sadly, there are other problems. Carrying hydrogen around in liquid form is extremely expensive, difficult, and dangerous. It is liquid only at −271°C. Liquefaction wastes 30% of the available energy. Things are not much better using the gaseous form. Hydrogen gas can leak through metal. In an enclosed space, at modest concentrations, it needs very little help in ignition. The tiniest spark will do. Worse, the flame is invisible in daylight — witness the tail of the Space Shuttle at lift-off. NASA is said to have once recommended walking with a broom extended ahead as a fire detector. Hydrogen also has the potential to explode in confined spaces.

It is, however, a mistake to condemn hydrogen as unsafe. At low concentrations, it is no more easily lit than methane. It has to get much hotter before it spontaneously ignites, and then it burns at only a slightly higher temperature. Being far lighter than air, especially when hot, burning hydrogen lifts rapidly. Burning quickly, it raises less risk of secondary fire. Finally, combustion produces only harmless water.

The spectre of the Hindenburg can be confidently lifted. That fire was caused by a highly flammable covering material. Most people died by falling or jumping. Only two were burned to death, most likely by the craft's skin or by the diesel stored aboard to fuel the engines.

Technology and mass production will undoubtedly resolve the issue of cost. For hydrogen, the problem will remain one of supply.

By definition, an automobile carries its fuel supply along with it. This has been an obvious disadvantage right back to the *Pennydarren*, as has been noted already. All anyone really wished to propel was the load, be it animal, vegetable, or mineral. Lugging fuel as well was not easy to justify. As well as the clear inefficiency — even after a hundred years of development, a steam locomotive might still haul an equal weight in coal and passengers — it introduced the problem of fuel distribution.

An advantage of rail over road transport, which has been clear for a full century, is that fuel can be offloaded. All around the developed world, railways have been converted to use electricity, transmitted from a fixed source. The only fly in the ointment is that the transmission of electrical power is 'lossy'. As much as half the energy generated may be lost in the process. Against this must be weighed the inefficiency in both carrying fuel alongside the load, and its independent carriage to depots and bunkers.

Electricity can also be generated remotely and thus potentially both cleanly and cheaply. The French rely heavily on nuclear power to generate electrical power for their *Train à Grand Vitesse* (TGV). However, the decision as to how electricity is generated is rendered *separate*, and can be taken according to considerations such as security of supply, sustainability of supply, and environmental impact.

There is no such liberty with road transport.

This is a good point to ask about energy efficiency with regard to road and rail. We now ask, with any vehicle, where does the energy go?

It might at first seem that most of it would be required simply to overcome 'inertia'. As pointed out long ago by Sir Isaac Newton, an object remains at rest, or in uniform motion, until acted upon by a force. This is often known as the "law of inertia" and is properly attributed to Galileo. If you want something to move then you must overcome its inherent inertia by applying and maintaining a force. In doing so, you will consume energy in proportion to the force itself and the period of time in which you maintain it.

Once the object is moving, it is said to possess *kinetic* energy.

The problem for transport is not so much building up kinetic energy as keeping it. Frequent stops are what makes the bus so inefficient (as well as slow). The same problem affects driving a car around town, though to a lesser extent as the stops are less frequent and the vehicle less massive.

It might seem surprising, but the energy lost on each stop, given a typical car and a typical speed around town, is only about 30Wh.

Any vehicle can recover at least half of the energy it loses this way by using *regenerative* braking, which we met in Chapter 2.

A bit more of a concern for smaller and lighter vehicles is the need to overcome rolling resistance. The first rails were laid for precisely this reason. Wheels run much more easily on rails than on any road.

Rolling resistance is proportional to the weight of the vehicle. The actual value varies according to the smoothness of wheel and track. Since the energy consumed (or "work done") overcoming a force equals the product of force and distance, the power consumed is just that of force and velocity.

Perhaps surprisingly, this turns out to be more significant, especially at low speeds. Around town, expect to use about 150W continuously. Since consumption varies linearly with speed, double this for the highway.

Overcoming rolling resistance is exactly like going up an incline. For rubber tyres on a road, the equivalent gradient is about 0.01 (one in a hundred). To work out the power used travelling up anything steeper, just multiply the above figure by the appropriate ratio; then halve it, to allow for the free ride down again (assuming it's not a one way trip uphill).

All these calculations assume a vehicle mass of one tonne, so if you live in a hilly area, you might see the benefit in something a bit smaller. Every contribution considered so far would diminish linearly with mass.

A one-tonne vehicle travelling faster than about 30kph is actually affected most of all by *air* resistance. It might seem again surprising that the speed at which this comes to dominate power consumption is so low. While there is something to gain through aerodynamics, as the lovely form of a Ferrari demonstrates, the greatest variation by far is according to speed.

The typical family car must continuously generate over fourteen *kiloWatts* to maintain 100kph. At 140kph, this rises to nearly forty.

While the power needed to overcome inertia and rolling resistance varies in direct proportion with the final velocity, that needed against air resistance increases with velocity *cubed*.

Of course, a higher speed does mean you get there more quickly, which reduces energy consumption in proportion to speed, but this still leaves a dramatic dependency. A 20% reduction in speed would save more than a *third* of the energy consumed on the motorway.

For a train, things are different. Because the weight is much greater, rolling resistance remains dominant up to a much greater velocity — around 120kph, for a typical inter-city express. The total energy consumed is therefore somewhat less dependent on speed, until it rises to a much higher level.

A train that devotes the same mass to each passenger seat will also see an advantage at suburban speeds, due to the much lower rolling resistance.[7] An iron wheel rolling on an iron rail offers only a fifth the resistance, reducing energy consumption in direct proportion.

We still need to keep an eye on the kinetic energy lost, however. Heavier trains should stop less often, and possess regenerative braking.

People are quite heavy (and getting heavier). It is enough of a problem moving *them*, without adding any more mass than necessary to the vehicle. The same applies to moving freight. When an engine, transmission, and fuel, are added, the chassis must be made stronger, and thus heavier.

Here, we come to one of the principal problems in transport.

We need a vehicle with as low a "mass ratio" as possible, if we care at all about energy consumption. Let us assume for now that we do.

To keep the numbers easy, let us suppose a typical passenger weighs 100kgs. The car in my garage has a maximum payload of 525kgs and an unladen weight (mass) of 1,255kgs, giving a best mass ratio of about 3.4. But cars are rarely driven fully laden, as we saw in Chapter 2. If we instead assume a typical average occupancy of 1.2, then we obtain a more realistic figure of 11.5. In other words, when I venture forth in my chariot, I am actually moving around more than *ten times* the mass I wish to carry.

As a result, the true energy efficiency of my car is less than 2%.

---

7 Exasperated by the repeated claim that buses are cheaper to run, a private train operator once swapped the tyred wheels of a bus for flanged iron to demonstrate how much further it could run on the same measure of fuel.

I could have bought a cheaper, smaller, lighter car. Like most people, I like to have a little power in reserve, though not as much as I used to enjoy in my misspent youth. I know that laboured engines last less long, and that sometimes one can get out of trouble with a little extra shove. I also have a family and may occasionally wish to fully load my vehicle. The engine is capable of developing 122bhp (about 90kW). A third of that power, and thus a much lighter engine, would suffice most of the time.

By far the greatest contribution to the mass of my car can be attributed to the attempt to render it safe. My children are in there sometimes. Having previously suffered at least three life-threatening road accidents, I know the threat. The thing is made of steel and has two body shells with a very strong design. Finally, as any mechanical engineer knows only too well, once the basic structure gets bigger and heavier, everything else grows in proportion. The heavier the vehicle, the heavier the engine, suspension, *etc.*

In Chapter 2, we noted how dangerous the railway was in its early days and how public horror at each catastrophe led to it eventually being rendered safe. Recall that safety came at a price. Railway carriages have since been built to withstand high-speed collisions. As a result, the mass ratio of a train increased to eight or more, with 50% occupancy, and is thus comparable to that of a family car. Rail appears only moderately competitive with regard to energy efficiency because it has been rendered *safe*.

This is particularly true when contrasting bus and train. No bus will survive an argument with a railway carriage. Safety is the essential difference between the two. Without enforcing a similar standard for both in this regard, no comparison between their energy efficiency would be fair.

Nor can there be fair competition.

One cannot draw a line under the issue of energy efficiency on the road without mentioning the "sports-utility vehicle" (SUV). In 2003, nearly half of all new car sales in the US were SUVs, which typically possess twice the mass of a large family car and three times that of a small one.

The trend hopped the pond. I can see two from my window right now. Perhaps my neighbours, like many farmers, building contractors, *etc.*, have a legitimate need for one, but then many can be seen used for the morning school run. Their mass ratio, for such purposes, exceeds 25. I used to go to school on a bike (mass ratio less than 2) or walk (mass ratio 1).

Some owners have been honest enough to admit that their choice was partly to defeat speed bumps. Some bought one because they feel safer.

There is evidence that the occupant of a SUV is more likely to survive a collision. Unfortunately, the other party, unless similarly protected, is *less* likely to survive, especially if a pedestrian, and especially when a child. Overall, the risk to all concerned rises. Not only does increasing the mass of the vehicle worsen energy efficiency, it actually *reduces* safety.

Even before accounting for excessive mass ratio, the car is horrendously inefficient. An SUV reduces energy efficiency to a figure below 1%.

It is hard to imagine a more wasteful form of transport.

Or, for that matter, a more dangerous one.

To account fully for energy consumption, we ought also to consider the production and distribution of fuel, and the manufacture and disposal of the vehicle. The three costs are sometimes called 'well-to-tank', 'tank-to-wheels', and 'cradle-to-grave'. Only the second have we thus far assessed. The first causes about 15% of the total. Consumption brought about through the lifetime of the vehicle makes up only around 10%.

Once again, the car proves horribly inefficient, with a design life of only around 150,000kms, over just ten years. Buses are no better. Virtually no recycling takes place, and there is no "closed cycle" manufacture.

Rail at least offers longevity. Trains are typically expected to cover 5–10,000,000kms and survive *three times* as long. One British fleet of express trains is being refurbished, as I write, after thirty years of service.

Overall, rail offers between a hundred and a thousand times better cradle-to-grave consumption than road transport. Electrification allows the same high distribution efficiency as it does with traction.

Air travel is undoubtedly a wonderful thing in that so many of us can enjoy different cultures and the faraway places our ancestors only visited in war. With regard to our "energy security", however, its popularity poses something of a problem. Flying consumes more than *ten times* the energy used when travelling the same distance (overland) by train.

While long-haul aviation has risen slowly, the growth of *domestic* air travel is staggering — up 77% in a single decade in Britain, which typifies the trend worldwide. While the industry can claim a small improvement in energy efficiency, this is mainly down to filling a higher proportion of seats on each plane as a consequence of rising demand. The downside is an increase in (single source) energy consumption by more than half.

No one sensible wishes to see air travel returning to the preserve of the affluent élite. Many are the hypocrites who voice loud objections and then quietly board a plane for their holiday, or excuse themselves business travel on grounds of self-importance. I, for one, would not trade our family holidays in Florida for all the tea in China. Nor would I wish to see my children grow up with the same petty national and racial prejudices that remained common among the untravelled elders in my youth.

One ray of hope is the introduction of even bigger jets for long-haul flights. While these might seem a step in the wrong direction, the intent is to reduce the fuel burned per passenger. The Airbus A380 is the most visible attempt, carrying up to six hundred passengers, and reducing kerosene consumption to less than three litres per seat per hundred kilometres.

As demand rises, such giants become attractive hub-to-hub. Many more destinations warrant direct service, reducing the actual distance flown.

None of this affects the fact that, while rail offers far the best, and most improved, energy efficiency for journeys at the high end of our domain of interest (10–1,000kms), aviation has eaten up all the savings, and more besides. Ever rising demand for air travel may similarly deny any savings due to more efficient aircraft design.

All is not yet lost. The jet engine does not necessarily depend upon fossil fuel. Something similar to kerosene can be derived from biomass, and there is no insurmountable obstacle to burning hydrogen, given an adequate source of supply.

The first experiments, running a jet engine on hydrogen, took place as long ago as 1956 in the USA. A collaborative project between Russia and the EU called 'Cryoplane' commenced in the early 1990s, modifying a conventional airframe to demonstrate a hydrogen-powered airliner.

While only minor modifications are needed to allow a jet engine to burn hydrogen, tanks and fuel management systems are a different matter entirely. A far bigger problem is the fourfold greater fuel volume, which reduces aerodynamic efficiency and speed. On the other hand, the much greater energy density of hydrogen implies less energy wasted carrying fuel aloft. Then there is the problem of an adequate fuel supply.

Long-haul aviation may have a future, though at a higher price.

The current option to fly overland distances of less than 1,000kms merely reflects a low energy price that takes no account of resource depletion. Once the price of fuel rises to reflect a secure and sustainable supply, short-haul aviation is unlikely to remain competitive with modern electrified rail, which can offer similar journey times, door-to-door.

Summary

Energy efficiency varies widely with transport mode. Table 4.1 offers a summary for Britain, including change over the past decade. Sadly, that change favours modes which offer the least, and which are improving the least.

We have disregarded the bus here, for all the reasons discussed in Chapter 2. Bus transport is simply not competitive, with regard to any of the parameters identified therein. Buses are slow, polluting, and expensive, when their track cost is properly counted. While safe enough to ride on, they also pose a serious threat to both cyclist and pedestrian.

Road travel dominates transport, for short journeys at least, though rail travel is growing rapidly now the industry is at last receiving some investment after a century of neglect. Rail is still preferred for inter-city travel, especially for commuting to a city like London, which now imposes very high parking charges and even a "congestion charge" for entering the centre. Over a decade, rail operators have raised energy efficiency by 58%, despite beginning with a figure already over *three times* that of road.

In comparison, the automobile industry has achieved only negligible progress, partly because of the rise of larger and heavier vehicles.

At the end of the day, the facts regarding energy and transport are fairly simple. Transport currently accounts for over a *quarter* of our total energy consumption, and over two thirds of our oil consumption. The latter is so far beyond our ability to supply locally that it poses a significant threat to our security, as we shall see.

While "energy intensity" (energy consumed per unit of production) is falling, transport is actually increasing our overall energy demand. Road transport accounts for ten times more travel than rail, but is nearly eight

mode	year	usage *bn pass-kms*	growth %	energy *mTOE* [1]	growth %	efficiency *Wh/ pass-km*	growth %
road	1992	630		39		720	
	2002	730	+16	42	+8	670	-7
rail	1992	38		0.72		220	
	2002	48	+26	0.38	-47	90	-58
air	1992	4.8		0.52		1,260	
(domestic)	2002	8.5	+77	0.80	+54	1,090	-13
bike	1992	5		0		0	
	2002	4	-20	0	0	0	0
total	1992	678		40		689	
	2002	791	+17	43	+7	634	-8

TABLE 4.1 Usage and energy efficiency by transport mode (UK Dept. for Transport).

[1] Megaton of oil equivalent. 1 MTOE is typically equated with 11.6 tWh.

times less efficient. It is thus responsible for four fifths of transport energy demand. Meanwhile, far the most efficient transport modes of all — walking and cycling — are in rapid decline.

Despite an upsurge in bike ownership, cycling is in free fall — down 20% in Britain in a single decade. As we concluded in Chapter 2, the reason is simple — the threat posed by rising motor traffic. In the absence of properly separated cycle-ways, few now consider it a safe option.

There is plenty of evidence that investment can turn all this around.

When the railway receives attention, both usage and efficiency improve dramatically. Electrification is one example, and can afford access to diverse sources of energy supply. High-speed lines can compete effectively with short-haul aviation, as has been demonstrated in France. Britain, at last, has plans for further electrification and new high-speed lines, but lags decades behind the rest of Western Europe. In the US, inter-city distances are typically much greater, but many opportunities still exist.

Cycling has also been repeatedly shown to respond well to investment. Despite the small sums required, and the huge payoff in energy saved, reduced congestion, and improved health, spending remains negligible.

A failure to invest in alternatives to inefficient fossil-fueled transport, will leave us severely dependent on a single source of energy supply.

Oil soon be gone

We do not live in one world but two, sometimes referred to as the 'developed' and the 'undeveloped'. Sometimes people speak of three but let's keep it simple and call these worlds simply the 'rich' and the 'poor'.

Only around a fifth of the global population live in the rich world. Half have no access to commercial energy at all, and burn only biomass (mostly wood). They account for just a tenth of the world's energy consumption.

A global average per capita power consumption (about 2.2kW) is thus rather meaningless. In the poor world, the figure is only around 0.5kW, while Western Europeans consume more than eight times as much (4.2kW), and Americans more than two and a half times as much again (almost 11kW).

There is nothing wrong in principle with consuming lots of energy. In much 'green' literature, it seems to be assumed that it is inherently a sin. Either way, most would likely agree that consumption should be *sustainable* and that the full cost be met by the consumer. While the consumer may be happy if they are paying less than the full cost, someone else is surely not. If the current supply cannot be sustained then the consumer will soon be compelled either to change to one that can or to cease consuming altogether. We shall therefore distinguish sustainable supplies from depleting ones.

Whether or not nuclear power can be considered sustainable, or even an acceptable candidate, will be left open for now.

Almost *all* energy consumed in the rich world is derived from fossil fuel. In the USA, 40% is from oil, 23% from coal, and 21% natural gas. Half the rest is nuclear, despite the fact that no new nuclear power station has been commissioned since 1979, when a disaster was narrowly averted.[8]

The largest sustainable supply is hydro-electric, contributing just 3%.

A full quarter of the energy consumed in the rich world is used for transport, four fifths of which is by road. Currently, this is almost the exclusive province of oil, with only a tiny contribution from biomass.

The USA alone gets through almost a quarter of the world's total oil production — around a gigatonne[9] yearly, equivalent to 7.3 billion barrels.[10]

---

8 This did not prevent construction of fifty-three reactors already commissioned.

9 A gigatonne (gt) is a billion tonnes. A tonne is a thousand kilogrammes (kg).

10 A barrel holds 42 US gallons (≈ 35 Imperial gallons, or 159 litres) and weighs 136kg.

Almost half goes to the gasoline pump. A fifth is refined into fuel for heating. The rest includes asphalt and a variety of industrial feedstocks.

While America's reserves of coal are plentiful, her oil reserves would now last less than four years. She already imports between half and two thirds of her oil. The primary problem is the rate at which global reserves are being drained. Put simply, the world gets through over thirty billion barrels a year. Known exploitable reserves of crude oil run to between 1,700 and 2,250 billion, depending on who you believe. New discoveries may sound big but they pall when set against such voracious demand.

It will all be over in sixty to eighty years as things are going.

A typical economist would argue that *any* shortfall in supply will be corrected by market forces alone. When demand exceeds supply, the price rises until a new supply becomes 'economic'. No distinction is made between sustainable and depleting resources. Demand can exponentially rise, presumably to infinity, without any limitation being encountered.

Economists appear to believe in magic. Any warning, such as that issued thirty years ago by the brilliant Ernst Schumacher in his book *Small is Beautiful*, is dismissed as cynical or somehow 'extreme'.

The situation is actually much worse than a cursory glance at resource and consumption makes it appear. Demand in the rich world is rising rapidly. We have already seen, in Table 4.1, how British air travel has risen 77% in ten years, consuming 54% more fuel. Despite some of the world's worst congestion, and a 26% rise in rail travel, road transport has still risen 16% in the same interval, while cycling has declined by 20%.

These trends are typical in the rich world. As mentioned earlier, in 2003 nearly half of new US car sales were of SUVs, which consume around twice the fuel for the same journey. Efficiency is just not a consumer issue while gas is cheaper than cola. The market price for energy fails to reflect resource depletion, and so demand grows without restraint.

Presumably, we will continue pumping faster until supply fails.

The US Government signals "no problem" with fuel taxed at an average of just 27% — less than half that typical in Europe — and even gave a tax deduction to those who bought the largest vehicles for business use. American oil industry continues to benefit from a range of tax breaks, from drilling costs to a free clean-up of pollution. There has even been a tax *deduction* for the depleting nature of their product, where one might expect a hike.

It is indeed hard to understand why any government would reward a company for draining faster a finite resource on which the nation has become completely dependent.

Now for a third problem. There are new kids on the block and they want to play too. China already burns nearly two billion barrels a year despite reserves of only eighteen. It gets through as much as Japan and is expected to rival America by 2025, supping around ten billion barrels a year. In 2005, the Chinese bought two thirds more cars than they did in 2002. We have cars: they want them too. Second-hand bikes will be cheap.

Rising demand brings the horizon closer. Exponentially rising demand brings it *much* closer. Estimates typically predict exhaustion now in perhaps just thirty years. There is no economist's magic to let us off lightly with slightly higher prices. All the exploration has been done. The world is now thoroughly mapped. All the big deposits are known. The little ones will only forestall execution by a few years at most.

Sadly, we're not through yet.

There is a fourth problem and it's a real lulu (whatever a lulu is).

Oil is not distributed evenly around the globe. It comes in a small number of big pools. But that is just a nuisance. Much more troublesome is *where* these big pools happen to be. About half of them are in the Middle East — dominated by Saudi Arabia with about two hundred and fifty billion barrels, followed by Iraq, United Arab Emirates, Kuwait and Iran, each with around a hundred — not the most politically stable part of our planet.

By a strange coincidence, the world's biggest oil consumer, whose entire economy is utterly dependent upon an adequate supply, invaded Iraq, on the pretext that it possessed weapons of mass destruction. These failed to materialize, as the UN Weapons Inspectorate said they would.

Chaos reigns for now in Iraq. It remains to be seen whether invasion and "régime change" promotes regional stability in the longer term.

Saddam was undoubtedly a threat to supply: he had little love for America. Next door, to the east, Iran is hardly fond of "the great Satan" either, and appears to be attempting the development of nuclear weapons.

Probably the biggest headache, for those whose job it is to keep the black goo flowing, is in Saudi Arabia. All they have to sell is oil. They can hardly welcome the prospect of pumping out the lot in just a few decades. Their oil infrastructure has also allegedly been neglected, so that output cannot easily be raised to meet forthcoming demand anyway, without outside help.

The Saudi régime has many internal enemies. There is unrest, extending from political dissidence to fundamentalist ambition. Inviting western (largely American) expertise back in to rebuild infrastructure and increase output might easily bring about revolution or, at least, civil war.

A Middle Eastern oil supply cannot sensibly be regarded as reliable.

Unfortunately, the situation is little better elsewhere.

Venezuela, sitting on about seventy-five billion barrels, is unstable. Its former-paratrooper president is sometimes too nervous to leave the country, having been removed once already by his own military.

He is openly hostile to the US.

Libya (thirty-six billion barrels) is stable but, shall we say, still a bit frosty. Nigeria (twenty-five billion barrels) is swarming with armed gangs, perfectly happy with the worst terrorist tactics. One of the most powerful, with paper-thin political decoration, has made it clear that oil company employees are legitimate prey for kidnap and ransom.

Russia (sixty billion barrels) is both stable and reasonably friendly. The extent to which she is willing to trade with her former *bête noir* remains to be seen. She has markets just as big closer to home — her own, and those in Europe and China.

Only one large reservoir remains. Canada has a vast area of tar sands, where oil has been exposed to the elements and lost most of its volatile constituents. This could liberate the equivalent of around a hundred and seventy-five billion barrels of crude *but* almost a third of the energy is lost in its recovery. Tar must be baked at a high temperature. It also has to be "strip mined". Both processes are extremely destructive of the environment — all to yield perhaps just five or six years world supply.

We're still not through.

A fifth problem, and perhaps most damning of all to a policy of almost total reliance upon fossil fuel for energy, is the damage done to our global environment and the planetary ecosphere.

When any hydrocarbon is burnt, water and carbon dioxide are formed and released into the air. Unlike the principal constituents of our atmosphere — nitrogen and oxygen — carbon dioxide absorbs, and re-emits, thermal radiation. It acts much like a blanket or the glass roof of a greenhouse.

The surface of the Earth may be expected to warm up. Calculation of the temperature rise we may expect is extremely difficult because the interaction between ocean and atmosphere is complex and not fully understood.

A grand experiment has begun whose outcome is uncertain, but which may well be sufficiently profound to threaten human civilization.

A full discussion is left for the next section.

Sometime this century, our dependence upon oil must end. It is only a matter of when. There is also the question of how.

Back in 1973, America backed Israel in a war with its Arab neighbours. The Arab members of the Organization of Petroleum Exporting Companies (OPEC) responded with an embargo. When, in 1978, Iran underwent its revolution and cut off supply, the market price of a barrel of crude shot up to \$70. Rationing followed in the rich world.

Economists learned the meaning of the term 'shock'.

Of course, oil is not the only fossil fuel. Natural gas is by far the most popular immediate alternative. It can often be found alongside oil, and appeals to oil companies who can exploit it in much the same way. Unfortunately, it suffers the same problems. Global reserves are comparable to those of oil — about 110 gigatonnes of oil equivalent (gTOE)[11] versus 118–150gt — are being consumed at about the same rate, and are similarly distributed.

We should expect the same geopolitical price volatility and shocks.

Gas is also significantly more difficult and expensive to move around. It has to be liquefied, at a very low temperature, for seaborne movement — a process that consumes a significant proportion of its energy content.

A "dash for gas" may relieve some of the problems with oil in the short term, but it in the end the same constraints will apply.

It is hardly likely that the nations of the world will form an orderly queue right up to the final trickle, and then, all with equal aplomb, turn to something else. History tells a different story, without exception. There *are*

11 1 TOE is typically equated with 7.33 barrels.

alternative, and sustainable, sources of energy but they have received little attention, and still less investment. In any case, you cannot simply unplug an economy, or a private lifestyle, and plug it in elsewhere.

To quit an addiction is hard but it gets harder the more you postpone.

False dawn

It is perhaps understandable that previous generations have placed so much faith in technology. Such faith can be traced back to the Industrial Revolution. To those who survived, new technology must have indeed seemed wonderful. The railway and steamship opened up the world. The new 'manufactories' produced so much even their labour force could afford their output. "King Coal" could power it all.

For most ordinary folk enduring the revolution, the experience was rather different. It was one of appalling deprivation and suffering. A great many did *not* survive. The staggering irony is that those who did also gained great faith in mechanization. This was because they survived only by finding a place in the new order, typically tending machines in one of the new and dreadful industrial towns. Blind to all that had been lost, many ordinary people perceived their salvation in yet *more* technology.

The utopian age of science fiction began with the Victorians. H. G. Wells, Rudyard Kipling, and Somerset Maugham, all wrote amazing stories of wonder. I grew up on the last of the utopian sci-fi. The stories of Arthur C. Clarke inspired me greatly. He and my father were both radar technicians at the same time, and on the same air force base in Cornwall, during the Second World War. Like Clarke, my father was bedazzled by the new inventions he had seen and handled. Sitting on the end of my bed, he would tell me of seeing an early television set and the first jet fighter.

His generation can easily be forgiven their folly. Only by excelling in technology had we won a terrible war. The consequences of losing it remain unthinkable. That the Nazis developed the same technology, and more besides, and thus could easily have won, is rarely given proper consideration. German technology culminated in the V2 rocket. Declared impossible by British scientists, it led directly to the Saturn V and a moon landing.

Clarke was set apart from those near whom he was so often shelved. First, his work was truly science fiction. J. B. Priestly once commended him, in that his stories *required* their setting, and did not simply translate a fairy tale into space. Clarke himself distinguished fiction from fantasy in its concern for the possible rather than the impossible. Most so-called sci-fi, whether book or movie, is actually just fantasy. Second, Clarke could write the fiction because he knew the science. As a utopian, he led rather than drove. While others inspired only nightmares, he presented *positive* images of the future that were to inspire exploration, physical and intellectual. Third, his was literature, sometimes of high order. With Clarke, technology serves only to answer the spirit of our kind and our need to explore and find answers to basic questions. Only in that sense did it offer us salvation.

It was never merely about "new toys".

In the TV shows of my childhood, science fantasy abounded. People could do anything, even travel through time. Most of all, I recall the worlds of Gerry Anderson, where *everything* was nuclear powered. Concern for the environment lay far off in the future. For example, in Thunderbirds, someone builds a nuclear tree-felling road-layer, which goes predictably out of control, ploughs through a small town, and demolishes an innocent dam.

My brother and I immediately tried to build one in Lego.

How could one forget Grandma's new nuclear cooker. She complained she just couldn't get the hang of the control rods.

Nobody laughed at the time.

It has always seemed a little strange that if a government thinks a new idea sounds appealing they throw money at it. The better it sounds, the more they throw. One might think that if it was such a good idea then it ought to attract *private* investment. The excuse is often that the risk is too great. It seems that politicians, if not taxpayers, are happy with any degree of risk.

That the biggest state spender should be the world champion of free enterprise seems stranger still. Between 1947 and 2000 the US Government spent *a hundred and forty-seven thousand million dollars* on the development of nuclear power. None of that money appears as a cost in its delivery.

As an author of a book like this, I have devoted considerable time to research, often trying to find an answer to a simple question. The simpler the question, the longer it always takes. After considerable effort, I have abandoned the attempt to find an honest answer as to the cost of electricity generated by nuclear fission (never mind fusion). Claims are made that it is about 2p/kWh, which is comparable with around 1.5p for a contemporary coal-fired power station. Several costs are missing from this quotation.

First is that of research and development, *which is inseparable from that of nuclear weapons.* One British plant had been operating for sixteen years before it supplied a single drop of 'juice' to the grid. It is now officially acknowledged that power generation originated as a front to cover the production of nuclear weapons. It is not a mere "point of view".

The second uncounted cost is that of decommissioning. Both British and American governments have attempted to leave this with the operators.

The British Government has twice bailed out British Energy, the nuclear operator. The company is in no position to fund such an expensive and risky exercise as decommissioning a nuclear reactor. Around the world, few reactors have yet been decommissioned. The true cost remains unknown.

One engineer, leading the decommissioning of a reactor at Dounreay, when asked what it was to cost in a television interview, replied simply: "Ask me when we've finished." Meanwhile, particles of plutonium were showing up on neighbouring beaches, and on the sea floor nearby.

We were always told such things were 'impossible'.

Britain's Nuclear Decommissioning Agency (NDA) raised the estimate for decommissioning the current twenty UK reactors by 25% in a single leap to £70bn. Given that a typical member of this fleet generated around 400mW continuously for forty years, this incurs a forward cost of around 2.5p/kWh alone. Admittedly, a modern plant can comfortably generate 1gW, but we should respect the sentiment of the Dounreay engineer. The figure quoted by the NDA is just an estimate, given the constraints of today. Nobody knows what a new plant will cost to decommission, in around 2045.

No one in business likes to operate under the shadow of a huge and unpredictable cost that occurs *after* all returns have been gained. In America, no application was made by any commercial operator to build new nuclear plant in over thirty years. If the return outweighed the risk, surely someone would have done so.

A third cost, also likely to fall on the poor old unsuspecting taxpayer, is the depositing of nuclear waste. If the cost of decommissioning is uncertain, that of the disposal of nuclear waste is doubly so, since no permanent scheme has yet been publicly accepted. Nobody will ever want a nuclear repository in their back yard.

The term 'disposal' seems hardly appropriate for something that must be carefully monitored and guarded for tens of thousands of years. Deep underground storage has been proposed, but the safety of such a strategy depends upon a guarantee of geological stability over perhaps a hundred thousand years. There is also security to consider, in a world rife with people who are undoubtedly willing to use nuclear material as a weapon, and with rising sophistication. Only governments could ever commit to this.

Once again, the cost is as yet unknown, and seems likely to be high.

Finally, there is insurance.

Nuclear accidents have occurred, at least one with appalling consequences. No commercial provider has ever or would ever insure a nuclear power plant. They understand that a risk is not just the probability of an accident, it is that multiplied by a measure of its consequences.

The problem crosses international borders and affects foreign relations. A risk acceptable to one nation may not be so to a neighbour. Risk is also something of which, in general, we share a declining tolerance. Ireland is still demanding the closure of British Nuclear Fuel operations on the coast opposite its Eastern seaboard. But that is also down to a fifth unaccounted cost — that of pollution, of which tolerance is also falling.

In May 2005, it emerged that nuclear fuel, dissolved in nitric acid, leaked from a British nuclear facility, creating a toxic pond the size of an Olympic swimming pool, containing enough plutonium to make twenty atom bombs. To quote the Liberal Democrat spokesman for the environment :

> "The history of nuclear energy in Britain has been one of accident, leak, and cover-up."

A question of trust enters in.

Insurance protects against risk. The risks run by nuclear power go well beyond what happened at Chernobyl. Fissile material must move between many sites. It is impossible to eliminate the risk of that material falling into the wrong hands. Clearly the British Government agrees, at some level. We now witness regular public exercises by the emergency services in practice for coping with the effects of a "dirty bomb". Guidance leaflets have been posted through our doors at home.

Any new nuclear power programme would exacerbate such risks. The only possible insurance is for the military to guard all material all of the time. It is only fair to ask who would pay for that, and who would pay compensation when an accident occurs, as it surely will, despite the most earnest assurances otherwise.

Not long after 9/11, I had occasion to visit a nuclear power station. It sits on the south-east coast — a rather large and obvious target. I asked whether it would withstand the impact of a fuel-laden airliner. No, it would not. But who could imagine anyone doing something like that?

We no longer need imagination.

While we bother to set up substantial security forces, can it ever make strategic sense to establish our generating capacity with the smallest set of the least diverse resources. They then require strategic defence.

Again, we can fairly ask who pays.

There is one last nail to drive into the nuclear coffin. Inscribed on it are stories untold of dreadful human misery. We seem to die of cancer more often now and rarely discover the cause.

As I write, I sit on the northern apex of a triangle containing three nuclear sites. As far back as 1987, the British Medical Journal confirmed that children in the vicinity suffer a far higher risk of leukaemia. The risk doubles for small children within ten kilometres of a site. Information regarding the incidence of other forms of cancer in the same children strongly suggest a more than 50% higher risk. There is also evidence in childhood mortality figures that the risk was increasing as the century drew to a close. Evidence has been published linking these tragedies to plutonium leakage, like the one at Dounreay.

Once again, that issue of trust arises. We were assured it was safe.

In 1954, a time of great hope, and faith in new technology, Lewis B. Strauss, Chairman of the US Atomic Energy Commission, told our parents that their "children will enjoy energy too cheap to meter". While we can still admire the vision, perhaps, the truth is very different. Even 2p/kWh, with no proper account of research and development, decommissioning, insurance, pollution, and 'disposal' of waste, is a third more than current alternatives.

A realistic price would be far higher.

In 2003, the British Government took the view that :

> "While nuclear is currently an important source of carbon free electricity, the current economics of nuclear power make it an unattractive option for new generating capacity and there are also important issues for nuclear waste to be resolved."

No doubt after much professional lobbying, on behalf of its proponents, 'uk.gov' inverted this position in 2006 and opened the door to fresh nuclear proposals, in a belated attempt to patch up a failing policy of carbon emission reduction. Suddenly, figures are published that appear encouraging, especially in the light of a tax on carbon emission.

Yet, they challenge credibility. The cost of a new 1gW nuclear plant is around £1.5bn, but the new Government figures allow just 3% as the decommissioning cost. That is just £45m, and must be compared with the same government's estimate of £3.5bn for decommissioning one of the old stations — nearly *seventy-eight* times as much!

Yet again, there is the issue of trust. Truth is consistent.

French "lignes à grande vitesse" (high-speed railways) are powered by electricity, most of which is nuclear generated. Hence, despite appearance to the contrary, nuclear power *is* indeed relevant to transport, just as long as you don't need to carry the fuel and generator with you.

A nuclear reactor is rather heavy. The fuel is even heavier.

One of the greatest advantages of rail is the ability to connect it to an electricity supply, and avoid carrying your source of power. This is impossible with road transport, although there is one way in which a nuclear reactor can power an automobile. Nuclear fission produces hydrogen, in significant quantities. This could fuel a car, and in the perfect manner — a point not lost on nuclear lobbyists. The problem is the sheer quantity required.

One proposal combines several aspects of new technology. First, you overcome people's fear and suspicion by building gigantic nuclear power stations hundreds of kilometres from any city, preferably in a desert. There you produce electricity and hydrogen. To avoid the otherwise inevitable losses of both on their distant journey, you employ something even Arthur Clarke and Gerry Anderson never foresaw.

A very strange consequence of Quantum Mechanics is that materials sometimes lose *all* their resistance to the flow of heat or electricity; they become 'superconducting'. Thus far, superconductors function only at *very* low temperatures, far below the freezing point of water. At the temperature of liquid hydrogen, it is relatively easy to achieve superconduction.

It has been proposed to solve two problems at once. A pipe carrying liquid hydrogen is insulated, in part, by a superconducting lining. The lining keeps the hydrogen a dense and relatively manageable liquid. At the same time the hydrogen keeps the lining a superconductor.

Even the taps that would emerge at the roadside have a holistic ring. When you park your fuel cell driven car, you can either top-up your hydrogen tank, or sell the electricity you can produce into the superconducting grid.

Perhaps you have your own private source of hydrogen.

Something of the bold vision of the 1950s comes through with this. You have to admire its audacity, if not its price, which will prove astronomic.

There is one more futuristic vision for nuclear power.

You may remember that fusion of small nucleons yields even more energy than fission of large ones. Controlled nuclear fusion is not yet practicable, never mind cost-effective. It is likely to be much cleaner than fission, with, for example, no spent fuel to reprocess or bury. But high-level waste will still result and will still require transport and disposal.

Enormous research and development bills will still need to be written off. Even then it is expected to be even more expensive than fission. And it still concentrates generation in a small number of vulnerable assets.

Again, if the promise is so great, why is there no private investment?

The real reason for the gigantic investment of our hard-earned taxes in nuclear power has nothing to do with the generation of electricity. As mentioned earlier, at least one nuclear plant in Britain operated for sixteen years before it supplied any electricity to the grid at all. In fact, we are now allowed to report openly that the purpose of the first British reactor, at Windscale (now Sellafield), in Cumbria, was to manufacture plutonium, and other elements, for the British nuclear weapons programme.

Back in the mid-1950s, the pressure to match the American capability to manufacture nuclear weapons, using relatively meagre resources, was immense. It led to the world's first serious nuclear accident. The reactor caught fire and spread radioactive material over England and Wales. A nearby town was saved only by the direction of the prevailing wind that day. Fifty years later, radioactive material remains detectable.

The same hills, in Cumbria, were later polluted by Chernobyl.

When Britain and America object to the nuclear power ambitions of other nations, it is because they know it means bombs as well.

Perhaps the proponents of nuclear power have never seen *Metropolis*, or perhaps they failed to get certain aspects of its message. Those with a degree in physics may aspire to some measure of its understanding. They might also earn enough to own a few shares in a nuclear power company. But unless we can envision a world where the entire population can understand and share such assets, there it must fall.

We cannot live without an understanding of, and some control over, the things around us on which we depend. What is tyranny if it is not a world you can neither understand nor control. It really doesn't matter whether the uniform of those who do understand, and can control, is coloured white or khaki. They diminish us just the same.

There are simpler, and less remote, ways in we can meet our needs.

Sunrise

Some of us are old enough to recall the shock of that image, taken by an Apollo astronaut, of Earth, hanging in the void — a little blue and white ball, delicate and vulnerable. For the first time, we could fully appreciate our planetary home, and how utterly dependent we are upon it.

Those astronauts had to be careful with their cameras. They had no atmosphere between them and the blazing star we call the Sun. It is not the gentle mistress of the balmy summer's day that we take it for. It is a very real place of unimaginable storm and violence. Had the astronauts been less lucky with the solar weather, they might not have survived their trip.

Every space mission has a critical technical problem that must be solved. Indeed, solving such problems would be an excellent justification for any space programme. To go to the Moon, you have to recycle air. To make a space station worthwhile, you must recycle water. It is simply too heavy to bring all you need up from the Earth below. To go to Mars, you must learn how to recycle air, water, *and* food. You must also find a practicable way of protecting the crew from solar storms.

Despite its violence, the output of our star is remarkably constant. Solar luminosity is an incomprehensible 390 billion teraWatts. The energy per unit area incident on the Earth is called the "Solar Constant", and has been consistently measured by more than one satellite instrument as 1.37kW/m^2. That is, for every square metre, facing directly toward the Sun, you might, in principle, collect enough to heat a typical room.

My favourite story relating to the power of sunlight is Arthur Clarke's *A Slight Case of Sunstroke*, where vengeance is wrought on an unpopular football referee using fifty thousand copies of a large match souvenir programme with a silvered back. A simple mirror of sufficient size can make a very effective directed energy weapon (DEW).

The Earth presents a small disc to the Sun, and intercepts only a very small proportion of its total output. Of the 174,000tW shining on Earth, almost a third is reflected back into space, though the proportion varies over its surface, and over time. Clouds reflect well, as does the sea.

About 120,000tW remains and is absorbed.

Because the surface area of a sphere is precisely four times that of the disc it presents to the Sun, the energy arriving per *planetary* square metre, before scattering and reflection, is only around 340W. Around a quarter of this is lost by reflection, mainly by clouds, leaving an average of 250W/m^2 (or 250mW/km^2). This is the *ambient* solar budget, not what would be available to collect on your rooftop. Another fifth is lost by absorption, partly by clouds, but mostly by water vapour, leaving just 185W/m^2 as our *direct* solar budget. This is an average over both time and latitude.

There is no reason why one should not take a flat receiver and keep it pointing directly at the sun. It is a fairly simple thing to do. For as long as the Sun is above the horizon, and there are no clouds in the way, you could intercept up to about 1.2kW/m^2. Near the equator, this would be almost half the day, leaving an average of around 600W — all completely sustainable and free, aside from the cost of the collector.

Any competent engineer will quickly point out that recovering a large proportion of direct solar energy, as either heat or electricity, is somewhat ambitious. Solar thermal systems average at about 40% efficiency, requiring a typical minimum 2m^2 collector, say in the Mediterranean, for domestic water heating. Photovoltaic systems use silicon semiconductor material to convert part of the solar spectrum into electrical current. They typically achieve 15–20% efficiency, though close to 40% has been demonstrated.

All energy consumption must be balanced against population density. Everyone may claim their own entitlement but this must vary according to how many live nearby. For a change, this might favour the poor world, which tends more towards a distributed rural population. Each square kilometre could perhaps recover as much as 60mW at ground level (about a third of direct, or a quarter of ambient, solar budget).

Britain has one of the highest population densities in the world. If we ignore local concentration (around London) this affords an allowance for each of Her Majesty's subjects of around 200kW. That is still at least *forty times* their consumption, which can hardly be considered conservative.

In America, each citizen can enjoy a God-given bounty of about 1.7mW. That an American consumes, on average, more than twice the energy enjoyed by a European need thus have no consequence whatever.

As we said earlier, to consume energy is not necessarily a sin.

Of the 120,000tW ambient solar budget, about a fifth is 'consumed' raising water from the ocean and makes up the largest proportion which could theoretically be recovered. Water often ends up in high places where it retains considerable gravitational 'potential' energy. It wants to come back down again and join all the other water trapped above the Earth's thin crust. Place a turbine in its path and you get hydro-electricity. It works just as well generating a few kilowatts in your local stream as it does in the Three Gorges Dam in China, where the output may be as much as 18gW.

Of course, it would be somewhat ambitious to attempt recovery of all 24,000tW converted from solar to hydro-gravitational energy. There has to be a sizable output to justify investment but that threshold is undoubtedly much lower than would interest big industry. Capital concentration tends to favour a small number of immense schemes. Partly because of this, hydropower remains greatly under-exploited. As of 2000, only 700gW of capacity is installed worldwide, though this is still responsible for around 17% of electricity generation. Production could economically grow fourfold.

A further 1–2% of the ambient solar budget is converted to wind, which can be recovered using a windmill. Like hydro, recovery is not rocket science, nor does it pollute. Again, it is applicable on all scales, from a few hundred Watts for a sail-boat up to 5mW with a single turbine. Opportunities for exploitation are even more uniformly distributed, though far less well exploited. Europe has taken a big lead in wind generation, particularly Denmark and Germany, with around 3 and 12gW installed already (2008). Potential output is almost unlimited, especially if turbines can be maintained at high altitude, where the bulk of wind energy is to be found. This is a technical challenge, but no fundamentally new technology is needed.

Some argue that alternate capacity must be maintained for times of low wind, but this, unlike the electricity, cannot be sustained. First, it will always be windy somewhere. As installation spreads across an entire continent, and in diverse locations, supply from somewhere is guaranteed. While a single mill might cease turning altogether, ten thousand scattered widely, including some offshore, may be expected to vary in output by less than 20%. Second, their energy can be stored in a number of ways, from electrochemistry (batteries), through air compression, to hydrogen electrolysis. Third, no one is suggesting that supply should depend *entirely* on wind.

Other solar options exist alongside independent non-solar sustainable sources such as geothermal and tidal power. Even if one chose to continue burning fossil and/or nuclear fuel, wind is well worth exploiting.

True to form, people complain at the new and unfamiliar, and there is surely no excuse for three hundred large turbines atop a well-loved beauty spot. But c.1800, Britain had around 10,000 windmills. Most were well loved and would even feature in paintings. In time, the new mills will blend into the landscape, as did the old ones.

Wind energy is clean, plentiful, and affordable, at around 4p/kWh.

As to what is and what is not affordable, consideration must be given to external costs. Dominant among these are the consequences of anticipated climate change. According to an economic review, commissioned by the British Government, and conducted by the eminent economist Sir Nicholas Stern, the associated external cost amounts to around $85 per tonne of carbon dioxide emitted. For a coal-fired power station, this equates to about 8p/kWh, denying any claim that coal represents 'cheap' energy.

Someone must meet this cost, at some point in the future.

Were it properly counted, wind energy, along with many other sustainable sources, would be seen as the more attractive investment.

The final sustainable supply, from the ambient solar budget, is biomass. As a fuel, it is as sustainable as life itself, with about 1% of the ambient solar budget converted into chemical (hydrocarbon) energy by photosynthesis. Maize, for example, is 60% carbohydrate, and sorghum 70%.[12]

Biomass can be converted to ethanol in a time-honoured fashion. 'Ethanol' is just an abbreviation of 'ethyl alcohol', which simply refers to the familiar alcohol with which some of us occasionally intoxicate ourselves. The process of *fermentation* may have been discovered almost as long ago as fire.

An internal combustion engine can burn ethanol with very little modification. Alternatively, ethanol can easily be reduced to hydrogen for use in a fuel cell. Then emissions can be liberated where they can do no harm.

A hectare (10,000m^2, or just under two and a half acres) of maize can produce 3,000 litres of so-called bio-ethanol. Where it can be grown, sugarcane can produce 8,000. Brazil has demonstrated the production of nearly two thirds of its automotive fuel from sugarcane. Waste can be recycled as fertilizer, or anaerobically digested to yield 'biogas' (mainly methane).

As we all know, there is no such thing as a "free lunch".

There are three problems with biomass as fuel. The first is that distillation itself consumes much of the energy in the crop. In the case of ethanol from corn, only about a quarter of the energy in the crop is recovered. With sugarcane, that rises to about seven eighths.

The second problem is environmental damage.

In Indonesia, for example, deforestation has undoubtedly increased in response to the new market for fuel crops. As a direct result, the orang-utan now faces a serious and immediate threat of extinction.

In America, the great prairie has all but disappeared, along with several hundred species that depended upon it. Only a few percent of the original hundred million hectares remains. Corn and soybean crops are now worth more than cattle, which are now fed indoors anyway, removed from their part in the ecological cycle.

---

12 The terms 'carbohydrate' and 'hydrocarbon' essentially refer to the same thing — any compound rich in carbon-hydrogen bonds,

Forest and prairie grass actually make better candidates for fuel crops: between 50% and 97% of the crop energy can potentially be recovered. Unfortunately, the technology to turn cellulose into ethanol is still under development. Otherwise, marginal land and crop waste could replace an eighth of the world's oil consumption.

Meanwhile, massive agricultural subsidies foster destruction.

The third problem is interference with our food supply. As I write, a substantial rise in food prices is being blamed in part on increasing demand for fuel crops. However, there are always two perspectives in any market — that of the producer, and that of the consumer. We must also remember that reliance upon fossil fuel implies an unsustainably low price for energy.

We must accept a higher price in future, whatever the source.

Most rich nations are finding their agriculture in decline. The consequences are rural impoverishment, unemployment, and depopulation. There is also the threat of deterioration in the land and countryside which cost so much to bring into production. Biomass has the potential to reinvigorate agriculture with new products and new markets. Crop wastes can be fermented and new crops grown explicitly for fuel.

Forest and woodland can see a new lease of life.

Over the centuries, Britain has been extensively deforested to make way for arable farming. Much of that land is now out of production. If half was devoted to fuel production, it could supply around 10 *billion* litres, equivalent to about a third the British consumption of gasoline.[13] However, a Government report concluded that, for Britain to grow even a tenth of its current energy demand, some top-grade arable land would need to be cultivated to that end, in direct competition with food production.

Farmers and small towns are already tasting the benefit in America, where soybean, mustard seed, and rapeseed, are already being converted into 'biodiesel', which can be used in conventional diesel engines without modification. Hundreds of companies and public organizations already take full advantage. The potential in the US is immense, given around forty times the land but only five times the population, even when the *per capita* demand is double that in Britain.

---

13 Britain consumes about 20 billion litres of gasoline, but ethanol has about a third less energy per litre.

Small scale production of biofuels can yield a low cost because they can be derived from crop waste. This is certainly worthwhile and has other environmental benefits, such as reducing effective greenhouse gas emission. (If the waste is allowed to rot, it gives off methane, which will trap around twenty times as much heat in the atmosphere as the carbon dioxide that results from burning it.)

On the other hand, no farmer will grow a crop specifically to produce energy unless it is more profitable than food production. Even with low farm-gate food prices, energy crops are currently uncompetitive. Fuel prices would have to rise dramatically to change that situation. In other words, were we to move from fossil fuel in our tank to biofuel, we can expect a significant price rise. Exactly how big a rise is a matter for debate.

It may depend on emerging technology.

Algae can be grown intensively to produce perhaps as much as 60,000 litres of biodiesel per hectare — more than *ten times* the productivity of sugarcane. Such methods can avoid interference with both environment and food production. But it will still cost more than simply refining oil.

Once again, the problem, for the foreseeable future, is cost.

Solar power does not require the concentration of capital and our consequent dependence upon large institutions outside our control. It affords the possibility of *independence* for the community, and possibly even the individual. Urban life entails the surrender of such privilege. The inverse also holds: dependence invites urbanization. Central provision is always cheaper when homes are packed tightly in neat little rows.

Against a background of conformity, individuality is threatened.

Songs have been written about this. Probably the best known is *Little Boxes* by Malvina Reynolds, which gloomily relates the little boxes "made of ticky-tacky", in which so many are housed, with those in which they are buried. As a child, I heard Jake Thackery sing this and add a footnote of his own — "all the architects and planners all save themselves a cottage on the hill". A little more light-hearted, and from the same period, is Viv Stanshall's *My Pink Half of the Drain-Pipe*.

Independence is in the blood, at least, of everyone who values their individuality. While it is perfectly possible to render people just as dependent on ambient solar energy (sunshine, water, and wind) as they are now on fossil fuels, it is equally possible for the very same people to seize the opportunity for a substantial measure of independence via direct solar power (thermal and photovoltaic). Arguably the most significant difference with solar power is that there *is* a choice. There is even an overall and overriding advantage to going it alone: there is less need for conversion and transport, which waste energy and diminish security of supply.

Even when the individual cannot recover solar energy directly, it can still be collected much closer to home. As we have noted before, community is built upon sharing amenities and resources. Where nothing is shared there can be no community. The smallest village, in rich world or poor, can share a wind turbine. Agriculture — the original foundation for sustained communal living — can be renewed in the cause of energy production, as we have seen. The hand that feeds you can be that of a peer and neighbour, rather than a remote master. Of course, no individual is ever completely independent. Neither is any community, or any nation.

Like so many things, it is a question of balance.

Liberty depends upon maintaining the opportunity to establish our own balance with as much latitude as possible. An argument that urban life is out of balance was eloquently made in the film *Koyaanisqatsi*, by Godfrey Reggio and Francis Ford Coppola. The title is a native American term meaning "life out of balance". Time lapse photography is used to great effect, leaving only vagrants looking vaguely sane.

Solar power may well encourage what might be termed 'ruralization', reversing the current trend of urbanization. People will need space to collect energy directly and grow energy crops. They need forest and woodland for material and energy, just as their ancestors did down the ages. It is much cheaper and easier to live near the resources on which you depend.

Far the greatest reduction in energy use and carbon emission may be found in eliminating the *need* for transport, not in making it more efficient.

Ultimately, it may well be possible to put all the 'goodies' we enjoy on a sustainable basis and at the same time regain some control. What is required is not necessarily higher technology but more *appropriate* technology.

That which can sustain the rich world might also develop the poor one, and restore both our communality and security at the same time.

So what?

Automobiles are the least energy efficient form of transport. Limitations inherent in the use of an ICE, together with a mass ratio typically in excess of ten, lead to a figure usually below 2%. Excessive reliance upon road transport has contributed heavily to the lack of energy security now suffered by the rich world, particularly Europe and North America.

Rising consumption and competition over dwindling oil supplies means that this situation that can only grow worse. Increasing the rate of supply can mitigate price rise in the short term, but will precipitate a more damaging "price shock" in the longer one, as will the inevitable political turmoil.

As we shall see in the next section, the need to all but eradicate atmospheric carbon dioxide emission, especially in the rich world, is set to turn an already serious problem into a global crisis. Our complete reliance upon fossil fuel must now end, although carbon capture and storage (CCS) offers some hope for coal. It remains very inefficient and as yet unproven.

Nuclear power can make a significant contribution, but this would require the resolution of its problems of pollution, decommissioning, waste disposal, and risk of serious accident. Even if solutions can be found, that are acceptable to the general public, it is unlikely to leave nuclear power competitive with alternatives that are more sustainable.

An array of methods exist for the recovery of solar energy, which alone is easily able to meet global demand. Hydro and wind generation are now established but leave a vast potential for growth in coming decades.

Sustainable sources, including solar power, are well suited to electricity generation and thus rail transport, as well as domestic heating. They offer fewer options for road, with its requirement for portable energy. Both options for the solar derivation of fuel — biomass and electrolytic hydrogen — are problematic and likely to prove far more expensive than petrol.

Batteries remain heavy, limit range, and are too slow to charge.

Agricultural production of 'biofuel' is already causing severe environmental damage and will inevitably interfere with our food supply. It can also only ever answer a small proportion of current demand. Novel biotechnology (via cultivated algae) offers hope, but does not yet represent a practical and economic option. The inefficiency (and local noise and air pollution) arising from the ICE would, in any case, remain.

Solar-electrolytic generation of hydrogen may ultimately offer the only sustainable source which may prove adequate. However, the development of safe handling and distribution presents a serious technical challenge.

The electric motor has always had the potential to vastly improve both the efficiency and the design of the automobile. The future of the car now seems to depend upon the development of an affordable hydrogen fuel cell.

That constitutes a second substantial technical obstacle.

4.2 The burning question

Pull up another blanket

A quarter of a century ago, I found myself an unsuspecting cold-war warrior. I had the rather grand title of Engineering Physicist with a defence contractor. Working for the military industry while espousing pacifism is a fairly typical conflict of which I'm capable. I'm also a vegetarian who doesn't like vegetables. But hell, it was *interesting*.

Stratospheric sounding was my introduction to physics away from blackboard and book. As a species, we can now look down on our world from above and thus begin to appreciate its beauty and understand its climate.

With our technology, we can truly see the world as never before, but we can also bring harm as never before. My work was shared with other physicists with much greater experience. One found the time to help me understand 'our' instrument, and "radiative transfer" (the passage of thermal radiation) through an atmosphere. This was a critical influence in my life.

It also served to introduce me to the work of Sir John Houghton CBE, then Professor of Atmospheric Physics at Oxford University, Fellow of the Royal Society (FRS), and President of the Royal Meteorological Society.

Houghton was later to become Director General of the UK Meteorological Office, and later still a key figure in the debate on global warming and climate change, as Chairman of the UK Royal Commission on Environmental Protection, and Co-chairman of the Scientific Assessment Working Group of the Intergovernmental Panel on Climate Change (IPCC), to whom the UN refers. There are few with equal claim to speak on the matter.

Fascination with that instrument, which measured temperature within the stratosphere, and frustration with a rather lowly rôle in its life, led me to escape the clutches of industry and return to academia. While a devotee of space exploration, I was already aware of how much could go wrong with space projects and how slowly they progressed. So I chose to join a group building ground-based instrumentation for astrophysical research. I spent the first six months just reading about basic astrophysics.

A little astrophysics turns out to be just what we need to take the first step towards understanding climate on our unique and lovely planet.

Imagine the Earth as a bald and barren ball warming itself in the sunshine. Though some is reflected back into space — about 30%, on average — sunshine is continuous. One might ask why the Earth does not carry on for ever getting warmer and warmer. The answer is that it begins to radiate itself. All things radiate more as their temperature rises. It stops rising when they emit as much as they receive.

This is called "radiative equilibrium".

In the 19th Century, we learned that thermal radiation, emitted by anything hot, and visible light were one and the same thing. They are both examples of travelling waves of electromagnetism, and differ only in their frequency, or wavelength.[14]

The outer shell of the Sun — the photosphere — is extremely hot. We know that because it glows yellow. In fact, the photosphere emits a spectrum — as we see in a rainbow, where water droplets have spread it out — but there is more yellow than anything else. Our planet, on the other hand, does not get hot enough to emit any light we can see. Like a domestic radiator, its radiation falls below the red and is thus known as 'infrared'.[15]

It might seem odd that a body as cool as the Earth can form any kind of balance with something as hot as the Sun. However, the Sun is a long way away, and the Earth receives only a tiny fraction of the energy it emits. From the solar perspective, the Earth is tiny and lies in just one direction. To effect a balance, the Earth can radiate in all directions. Its rotation helps it lose

---

14 Since the velocity of light is constant, in any given medium, frequency and wavelength are reciprocal measures. (Multiply them together and you obtain velocity.)

15 'Infra' just means 'below'. 'Ultra', in turn, means 'above', or 'beyond'.

heat and ensures an even temperature around its circumference. The land and sea that warm up during the day, can radiate and cool down by night.

While the visible radiation from the Sun is absorbed by atoms, such as those of hydrogen and oxygen on the surface of the sea, the thermal radiation from the Earth is emitted by molecules. Both atoms and molecules can act like little aerials, emitting in just the same way as a radio aerial emits radio waves. A molecule acts like an aerial because electrical charge is distributed unevenly around it, leaving a 'dipole'.

Vibration is enough to cause emission. Reception is simply the reverse. Infrared radiation is absorbed causing vibration, which we observe as heat. This is how we warm our hands in front of a fire.

If Earth was indeed a barren orb with no atmosphere then equilibrium with the Sun could be reached with a surface temperature below freezing, at around −6°C.[16,17,18] Though some bacteria might survive, it seems unlikely that life would ever have arisen.

Earth would be as dead as Mars and the Moon appear to be.

Now let us suppose we add an atmosphere, of the kind that can act like a blanket to warm our planet.

A blanket intercepts and absorbs some of the heat we radiate. If it is cooler than we are, it will radiate less onward than we would have done. It will also reflect back at us the radiation it did not absorb, and return some it did. We therefore lose less heat overall, and our bodies need burn less fuel in order to maintain a healthy temperature.

---

16 In other books, even some that are otherwise reputable, you may see a figure as low as −18°C for the temperature of an Earth without an atmosphere, or with one that is free of any "greenhouse gases". This is wrong because it fails to account for the loss of clouds, which reflect much sunlight. Without clouds, only about half as much is lost by reflection, leading to a warmer planet. This error is a nice illustration of just how easy it can be to overlook a factor and be misled. Climate is complicated.

17 Physicists use an *absolute* temperature scale where zero denotes a point where no radiation is seen at all, where the atoms or molecules concerned are frozen rigid. Each degree resolves the same difference as does one in Centigrade. With no atmosphere, the Earth's surface temperature would measure 267°K, where 'K' denotes 'Kelvin' after a renowned 19th Century physicist.

18 Without oceans, which reflect sunlight better than land, this would rise to about −3°C, assuming a dry and lifeless rocky surface, like that of the Moon.

A balance ('equilibrium') must still be achieved, both between ourselves and the blanket, and between the blanket and the room. Inanimate objects cannot always simply produce less heat in order to establish a balance. If they continue to generate as much heat, they will grow hotter and hotter until all they produce *can* escape through the blanket. A heated towel rail can serve to demonstrate this. With no towels, it should be just warm to the touch. Cover it with thick towels and return after a few minutes. You will find the outside of the towels at almost the same temperature as was the rail. The rail may now be hot enough to burn you.

We regulate our temperature, keeping it constant. The effect of a blanket is to allow us to do so while saving energy, perhaps to fight infection. A towel rail, on the other hand, typically lets its temperature vary while keeping its energy consumption the same. Its temperature will become stable once "energy out" equals "energy in".

Earth is like the towel rail.

It's just plugged into the Sun instead of the mains.

An atmosphere differs from a blanket or a towel in one respect: while it affects the way energy flows out, it allows energy to flow freely in. This is known now as the "greenhouse effect", because a garden greenhouse works in much the same way. An atmosphere acts like a glass roof.

Glass is transparent to sunlight, which therefore warms whatever lies beneath, which then emits thermal radiation. A proportion of this is absorbed by the glass, which itself then warms up and radiates both down and up. What lies inside the greenhouse sees the radiation from the glass in addition to that from the Sun. A balance outside the glass can be reached only when the temperature inside is significantly greater than it would otherwise be.

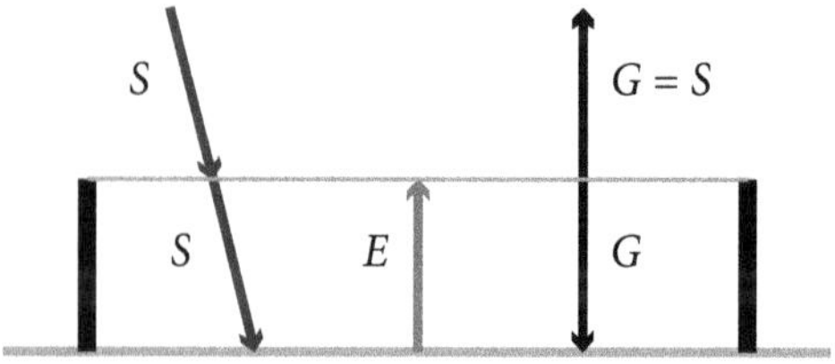

FIGURE 4.1 The simple greenhouse effect.

Figure 4.1 depicts an ideal system where the glass 1) is totally transparent to solar radiation, reflecting nothing, 2) absorbs *all* thermal radiation from below, and 3) is thin enough, or a sufficiently good conductor, that the temperature on top and bottom is the same. It will then radiate upwards and downwards in equal proportion.

Whatever lies beneath the glass then receives *twice* the energy it would otherwise because a balance ("radiative equilibrium") can be achieved only when the energy arriving is the same as that leaving.

Note the contrast between requirements 1 and 2. Glass just happens to have these properties. While it is famously transparent at wavelengths our eyes detect, it will pass only the 'near' infrared. (An *ideal* greenhouse glass would appear completely black at all thermal wavelengths.)

Given a balance, both inside and outside our greenhouse, we can perform a little mathematics and find that the ground receives, and thus must reradiate, twice the energy it would with no glass roof. Inside, E must equal $S + G$, but outside, S must also equal G. So E must be twice S.

Without the greenhouse, E would just equal S.

As I write, winter is setting in. One way to obtain a warmer bed is to use more than one blanket, as shown in Figure 4.2. Provided it can absorb all the thermal radiation from below, each extra blanket has the same effect as the first, requiring anything beneath to radiate as much again.

Temperature does not scale in quite the same way.

A single perfect blanket would lead to a thermometer rise of about 19%, but this relates to the *absolute* temperature scale, where zero refers to an object that radiates no heat at all.

Such an atmospheric blanket would warm a dry Earth's surface to around 48°C.

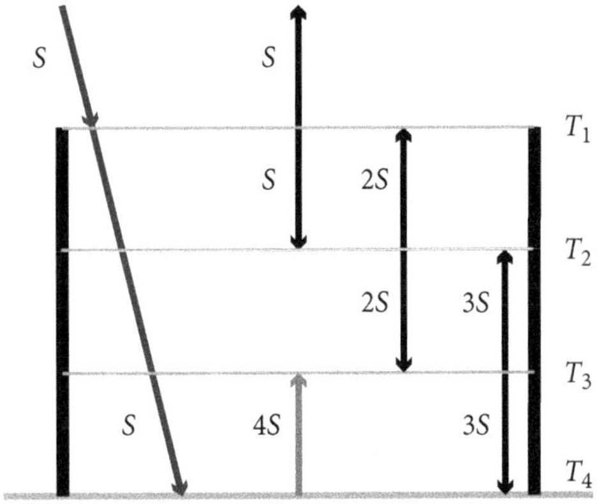

FIGURE 4.2 The compound greenhouse effect.

The number of effective blankets an atmosphere provides is called its "optical thickness". Earth currently has almost two. If the greenhouse effect were all that mattered, we would be forced to expect an average surface temperature in excess of 58°C, even after accounting for the loss of nearly a third of the Sun's incident energy by reflection back into space.[19]

19 This time, we do assume oceans and cloud; otherwise, the temperature rises to ~78°C.

Again, though some creatures might survive exposure to such an environment, life would be mainly restricted to mountain tops and the deep ocean. Some planetary explorers, who work thus far by robotic proxy, propose that life might arise floating in the sky. Temperatures do decline rapidly with altitude in our current model.

Perambulating apes would have a tough time indeed.

Things are a little more complicated still, thankfully. Air moves around when it's heated. It circulates, both horizontally and vertically, cooling both land and sea. The evaporation of water has the same effect. What we call 'weather' can account for the comfortable average of around 15°C that we actually enjoy.[20] Circulation and evaporation, as well as radiation, transport heat from the surface to the sky. While something on Earth has to radiate enough to create equilibrium.it does not have to be the surface alone.

We saw earlier that only half of the Sun's heat reaches the ground. While about 4% is then reflected, by land, sea, and ice, almost half, known as 'latent' heat, is given up in the evaporation of water. All that water must come down again, but not the heat. As it rises, the vapour cools and condenses back into liquid droplets. Just as evaporation absorbs heat, condensation releases it. The net effect of about 1m of evaporation and rain each year is the transport of a gigantic amount of heat from the surface into the atmosphere, at a wide range of altitudes. That heat is partly radiated out into space, restoring balance, but, at all but the lowest altitudes, the atmosphere is warmer than it would otherwise be. The surface is made much cooler.

The influence of clouds is far from simple. On their own, they reflect about a fifth of the incident solar energy. They also absorb and emit thermal radiation, and so add an extra blanket when they are around. (A cloudy night is usually warmer than it would otherwise be.) These two effects counteract, making clouds difficult to account for in any model. One cools the surface, the other warms it. The net result is believed to be a slight cooling.

Occasionally, and rather badly, I fly gliders — one reason why sheep are nervous in Oxfordshire. To fly a glider for more than a few minutes, you have to find a 'thermal' — a column of rising air, warmed by the ground

20 The term 'climate' usually refers to the overall *effect* of irradiance and weather.

beneath. What goes up comes down. Find the warm rising air, you go up. Find the cool falling air, you go down.[21] Experienced glider pilots can fly hundreds of miles in a single flight using just thermals, although there are other phenomena they can employ.

A thermal is an example of oddly named 'sensible' heat, collected from the ground by conduction and transported up into the sky by convection.

Again, the surface is cooled.

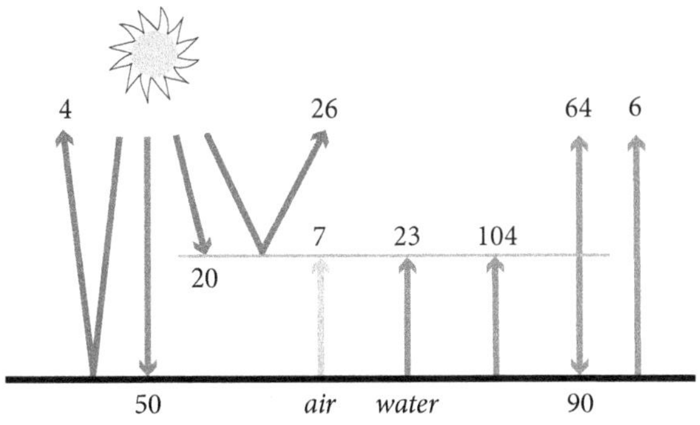

FIGURE 4.3 The solar energy budget of Earth, *c.* 2000.

Figure 4.3 depicts the solar energy budget of our planet and conveys something of the precarious balance on which life, particularly on land, depends.[22] Though it may seem absurd, our atmosphere is warmed more than twice as much by the Earth than the Earth is by the Sun. As well as absorbing a fifth of the solar supply, it gets the heat latent in evaporated water, in air brought by convection, and radiated directly from the ground below. Almost two thirds of this is returned to the ground by radiation. As a result, on the surface, we enjoy nearly *twice* the energy from our snug atmospheric blanket that we do from direct sunshine.

Such is the greenhouse effect.

A couple of years after my minor rôle in stratospheric sounding, I found myself sitting on a beautiful mountain in Tenerife nursing an instrument I had built myself. It was designed to look at sunlight reflected by interplanetary

21 Down-draughts can be powerful, and have been known to cause air accidents.

22 Planetary science like this would be impossible without space technology.

dust. One of my greatest problems, shared by many astronomers, was that the air through which starlight passed was itself glowing.

Airglow was a recent and unexpected discovery. Chemical reactions in the upper atmosphere cause the emission of visible light. Just as part of the solar spectrum falls below the red (infrared), some of it lies beyond the violet. Ultraviolet light has a high frequency and carries with it more energy — enough to knock electrons from atoms, ionizing them. Ions are capable of recombining to form new compounds. This is sometimes called "hot chemistry" and is believed to occur out in space. It seems capable of forming large molecules, even some precursors to life. In the upper atmosphere, nitrogen and oxygen are disassociated by sunlight during the day and recombine at night, emitting green light as they do so.

An atmosphere is neither simple nor inert.

Life is clearly very delicate. Without an atmosphere at all, our world would be too cold. With one, but without weather, it would be too hot. As it is, we are just about right. One might wax Darwinian and suggest that the only reason it *is* about right is that we have evolved precisely for that temperature. Were it warmer or cooler, we would have adapted and it would still be about right from the new perspective.

Unfortunately, now that we are able to go and look at some other worlds, we find no obvious signs of life in different environments. True, we have yet to look everywhere. Europa and Titan are still under investigation, but results thus far do not auger well. Most of those searching would be over the Moon (pun intended) to find microbes. As yet, we have no reason to believe that life, still less macrobiotic life, could adapt to any temperature much above or below that we are used to.

We should not confuse coping with diurnal or annual variation and coping with a significant change to the average ambient temperature. More than 10°C variation from the norm, and our very survival is threatened.

It may take a lot less to cost us our civilization.

We are dealing with an immensely complex system that we do not yet properly understand. Learning more about planets and atmospheres depends heavily upon our capability in space. We need to study more than one example. Complex systems *per se* remain a subject for research.

All that we have learned thus far suggests that we are much more fortunate, and much more vulnerable, than we ever suspected.

Our world is almost unique in the solar system in having retained a substantial atmosphere. That of other worlds has leaked away into space. Mars has but a remnant and is thus cold and desolate. Venus has retained hers, but more of that later. Earth also has entirely unique features, whose significance we are only beginning to comprehend. By far the most important, after the atmosphere, is the ocean.

Different materials require different amounts of heat to raise their temperature by the same amount. In other words, their ability to store heat varies. Water has a high "thermal capacity". Indeed, a swimming pool offers a very effective way to store heat in summer for winter. On the other hand, if you decide to warm one up in winter, you can expect a very large energy bill.

Living on land, we tend to forget just how much of the Earth is covered by ocean. It's more than half the surface. An ocean makes a wonderful transport system for heat, carrying it from the tropics to cooler latitudes, making them much warmer than they would otherwise be.

Though warm water will float above cool water, if for some reason it were to gain a higher salt content at the surface then it can become heavier and fall. When ice forms, or evaporation occurs, salt is left behind, increasing the density of the ocean surface. As a result, warm surface water at high latitudes sinks. A surface current is thus set up from low to high latitudes, known as the "thermohaline circulation".

Moving heat around the surface of the Earth makes more of it habitable and comfortable. Any change in the pattern of flow could have a disproportionate effect on some regions. Britain, for example, would be much colder without that part of the circuit known as the Gulf Stream. Life would be far less pleasant and productive.

Just like the atmosphere above, circulation occurs through all three dimensions. Oceanic circulation also moves heat from the surface to the deep, where a vast amount of heat can be stored, unknown to those above.

Any planetary warming might not become apparent for decades.

In describing the greenhouse effect earlier, we referred simply to the atmosphere absorbing thermal radiation. In fact, neither of its primary constituents — oxygen and nitrogen — do so. It is the *minor* constituents that turn air into a blanket.

Water vapour dominates, followed by carbon dioxide, methane, nitrous oxide, and ozone. A cocktail of large molecules produced entirely by human industry also participate. Though each has a strong greenhouse effect, they are as yet present in too low a concentration to play a great part.

We should be very glad of rain. As we have seen, evaporation of water is largely responsible for cooling the surface of Earth. There is another reason, however.

Now that our robotic progeny have visited it, we now know there is no Mekon on Venus, no Treens to attempt our conquest, no Therons to make good our waning food supply, and no belt of fire for Dan Dare to brave. Shame really.

Instead, there is an atmosphere a hundredfold thicker than our own, through which very little sunlight can penetrate. What is more, three quarters of the sunlight incident on Venus is reflected, compared with less than a third by Earth. Though one might expect the surface to be therefore cool, a temperature of 525°C has been recorded — hot enough to melt lead.

One sufficient reason for space exploration is to learn more about our own world by studying others. No amount of theory can ever substitute for observation. Venus and Earth are of similar size, share a similar position in the solar system, and are likely to share the same chemical composition. Having now looked at Venus, we might well ask why Earth has not already suffered a similar fate, and whether it could still do so.

We might begin by assuming the same starting conditions — each planet possessing an ocean but only a thin atmosphere. However, Venus is a little closer to the Sun and is thus in receipt of more solar energy. Evaporation would have been more rapid, and here we have the nub: water vapour is a very powerful greenhouse gas.

The greenhouse effect would further warm the surface encouraging more evaporation, and so on until either *all* the ocean has become vapour or the atmosphere has become 'saturated' (*i.e.* can hold no more).

In fact, the warm air of Venus would *never* have become saturated. A 'runaway' greenhouse effect would thus send its temperature soaring. On the cooler Earth, saturation occurred early on. No runaway effect is possible. Evaporation is free to continue cooling the surface. The result of any excessive evaporation is just rain. It can never rain on Venus.

After water vapour, carbon dioxide is the most important greenhouse gas on Earth, because of its high concentration. Since oxygen makes up 21% of the atmosphere and combines only too readily with other elements, we should be primarily concerned with carbon itself.[23]

About 750 gigatonnes (gt) of carbon are suspended in our atmosphere. This is small compared with other planetary reservoirs. Life and its residues on land store about 2,000gt. Fossil fuel deposits are therefore highly significant at 5–10,000gt, though by far the largest holder is the ocean with around 40,000gt.

We do not live on a still world. Carbon is regularly exchanged between air, land, and sea. In the 1950s, one Charles Keeling began measuring the carbon dioxide concentration in the clean skies above a mountain in Hawaii. He observed it rising and falling with the season.

Earth is alive and breathing.

Every year, I take up the challenge posed by my thirty metre hawthorn hedge. In a single year, it adds about 50% to its mass that I must either burn or dump somewhere to rot. The greater part of that material is carbon taken from the air and rendered solid. When burned or allowed to rot, it will all return to the air. To raise the energy for the job, I eat cereal and other carbohydrate, taking more carbon from the air, which I return on breathing out. The combination of photosynthesis, respiration, and decomposition, forms the *short* carbon cycle. It was this that Keeling observed. The carbon concentration would fall during spring and summer as crop and forest grew. In autumn and winter it would rise as their residue decayed.

The length of the short carbon cycle is dictated by biota lifetimes. As a result, carbon can complete the cycle in anything from one year to a century or two. A quarter of atmospheric carbon (about 185gt) moves in and out of the land and ocean reservoirs each year, though this represents a tiny proportion of either.

Not all the short carbon cycle is directly associated with life. Though the creatures of the ocean also consume and exhale, a large amount of carbon is also exchanged via chemical processes. Half as much again is exchanged between air and ocean as between air and land. While 90gt

23 Given that a carbon atom has twelve times the mass of one of hydrogen, we say it has an "atomic weight" of twelve. Oxygen is a little heavier at sixteen. One tonne of carbon therefore combines to form $44/12 = 3.67$ tonnes of carbon dioxide.

passes between atmosphere and ocean, this remains paltry compared with the vast amount the latter contains.

Recall that the *long* carbon cycle begins with volcanic eruptions, emitting vast quantities of carbon dioxide into the atmosphere. This can react with rainwater to form a mild acid which erodes rock containing calcium and silicon. Life on land can assist in such 'weathering'.

Continuously, over millions of years, calcium and silicon carbonates are washed down rivers and into the sea. Creatures there use these compounds to build shells. Although, following the death of their hosts, most of these decompose as they sink, a small proportion arrive on the ocean floor to form sediment. Over tens of millions of years, carbonaceous sediments are drawn underground (subduction). There, they may be changed by heating (metamorphosis) and later ejected back into the atmosphere by a volcano (eruption), to begin the cycle once more.

The quantities of carbon that flow along the long cycle are vastly greater than those that flow along the short one. Any accidental connection of the long path onto the short may be expected to precipitate a substantial change on the timescale of the short cycle.

After carbon dioxide, methane is the next most significant greenhouse gas, given off by decaying organic waste. Although the air naturally contains about four hundred times less methane than carbon dioxide, the former absorbs over seven times as much thermal radiation and thus has a much stronger greenhouse effect, tonne for tonne.

Human activity now generates almost three times the methane produced by natural processes. Photochemical reactions in the upper atmosphere destroy it after a decade or so. While this process takes care of natural production, it is overwhelmed by our industrial contribution. Atmospheric methane is rising by 1% each year — twice the rate of carbon dioxide.

Finally, and with more than twice the concentration it would have naturally, comes nitrous oxide. Unlike methane, it lasts over a century before being broken down. Little is yet known of its origin but its contribution as a greenhouse gas could become significant if its increase continues.

Suppose the planetary surface temperature fell for some reason. The rate of evaporation would then also fall as a consequence. Less evaporation means more heat retained at the surface, and thus a temperature *rise*.

This is an example of "negative feedback", which promotes stability.

Sometimes, "positive feedback" can occur, exacerbating a disturbance, promoting instability. There are dozens, perhaps hundreds, of "feedback loops", both positive and negative, that affect a planetary climate.

Feedback loops can interact. A disturbance in one can activate another. Such "complex systems" can be extremely difficult to analyse and understand, even when each feedback effect is itself well understood.

Unlike systems we currently engineer, complex systems cannot be tackled by reduction. You cannot break down the problem recursively into ever simpler components that can be understood first separately and later in unison. Complex systems tend to be *holistic* and defy such an approach.

Sometimes, even a minute change can have a dramatic effect. Stability may return only under profoundly different circumstances. Chaos is even possible, where a system follows a path that never retraces itself, and thus becomes utterly unpredictable.

If the weather is hard to predict, climate change is doubly so.

Our environment is a *very* complex system. It is also one of which we almost certainly have incomplete knowledge. We cannot expect to foretell the outcome of any significant change. Because we now know such systems to sometimes be sensitive to even an infinitesimal disturbance, we must consider *any* change significant.

Respect and caution are highly appropriate.

Now and then

Most stars appear to shine with remarkable constancy. Precise measurements of solar luminosity only extend back about thirty years but indicate a variation of less than one tenth of a percent. Extrapolations, correlating solar activity with output, suggest that up to a half percent variation may have occurred over the last three centuries. A cool period on Earth in the 17th Century may have been a consequence.

On the other hand, we have long known that the orbit of the Earth wobbles over the millennia. We can reliably calculate that, for the last million years or so, its eccentricity increased a little every 100,000 years, bringing the planet a fraction closer to the Sun twice a year.

Growth rings in tree and coral can reveal the climate of the recent past, as can pollen found in glacier, ice-cap, and sediment below lake and ocean. The sedimentary pollen record extends back around fifty thousand years. Beyond that, we can resort to fossils of microscopic sea creatures. Ice core samples can reveal the carbon dioxide (CO_2) concentration over at least four hundred thousand years.

The climate record tells us that average surface temperature does indeed increase when the Earth ventures nearer the Sun. In between warmer periods are "ice ages", when glaciers extend down to European latitudes. The last one began about 70,000, and ended just over 10,000, years ago.

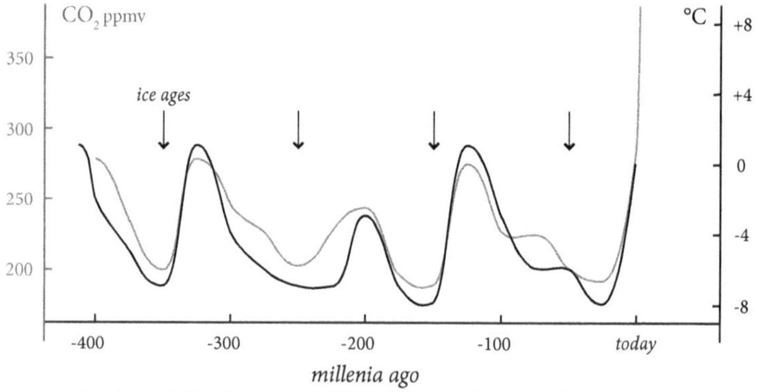

FIGURE 4.4 Temperature and CO_2 concentration over the last 400,000 years.

A very small change in solar input clearly brings about a *major* climatic response. Radiative equilibrium alone cannot explain such a dramatic reaction. Some positive feedback must play a part.

One effect arises because ice reflects sunlight very efficiently. As glaciers grow, more energy is lost in reflection, causing further growth, and so on.

If we overlay the records of surface temperature and atmospheric carbon dioxide concentration we see a strong correlation (Figure 4.4).[24] It

[24] CO_2 concentration is measured as parts per million by volume (ppmv).

is reasonable to believe that a change in the Earth's orbit caused a change in temperature, and that this in turn caused the variation in CO_2.

We expect a lower concentration to cause a further fall in temperature. Hence, the greenhouse effect is assumed to have provided further positive feedback, exaggerating the amplitude of the cycle. The importance of such feedback is borne out by attempts to reproduce the temperature change seen in the ancient record using computer models. These make plain the interacting feedback loops of a complex system.

With regard to ice ages over the last half million years, glaciation and an induced greenhouse effect are sufficient to explain the historical record.

Science is poorly understood by many people, including, sadly, many who call themselves 'scientists'.

There is nothing absolute about science. It can substitute for neither philosophy nor religion. We continue to suffer those who are knowledgable yet lack wisdom. Science seeks only to observe nature, and to explain many phenomena in terms of a minimum set of elements and principles. It can guide but not dictate, explain experience but not existence.

The human world has always rewarded those who claim to 'know', whether they indeed they do or do not. In the world of science, esteem is fought over with the same fervour as is money in the world of business. Truth and justice fall victim just the same, because the people are the same.

Neither grant nor publication follow an honest "I don't know."

Another pitfall is to mistake theory for 'fact'.

Science redefined the term 'knowledge' in a manner best described by the sign above the door of the world's first scientific society — The Royal Society, in London — which reduces to "Take nobody's word for it!" Scientific truth is accepted only when an observation can be repeated by any sceptic.

That is also how truth ceases to be absolute. Unlike religious 'truth', it is purely relative. Only the observation is accepted; interpretation remains open. Theory is accepted only by the measure of its predictive success. In the pursuit of science, experience shows the eventual failure of *every* theory.

Only through the failure of theory, has science ever progressed.

Here lies the final pitfall, in both the presence and nature of theory.

Science can be divided into three distinct camps, each of very different value in the confidence found in the knowledge they bring. Every branch of science begins with the empirical, consisting only of observation, with little or no theory. Sometimes, individual observations can be explained, but development has yet to bring a holistic account. Many branches of science remain at that stage today. In others, at least a partially integrated theory exists but is imprecise and usually merely verbal.

Only in the study of non-living physical phenomena has science approached its ultimate goal — a precise and integrated formal theory, capable of predicting the behaviour of most systems. Physics has given us both a satisfying account of the world around us and technology that has brought security and prosperity to all who can learn its wonders. Its domain extended to encompass chemistry, early in the 20th Century, and is now poised to envelop biology, and thus to bring a revolution in medicine.

The technology of life itself may become available.

Yet it has its limitations. One of those we have already met. Complex systems — those with multiple interacting feedback paths — are, as yet, poorly understood, and are, in any case, often intrinsically unpredictable. Some can alternate between predictable and unpredictable behaviour, which goes some way to explain why a weather forecast can prove reliable one day but not the next. Unfortunately, the switch itself is usually unpredictable.

We should be careful to distinguish observation from theory, and remember that neither observation nor theory is ever *wholly* reliable. The repeatability of any observation, and the reliability of every theory, must always be open to question, and accepted only by success on that basis.

Uncertainty, never certainty, is the stuff of science.[25]

One other item emerges from the paleoclimatic record.

While each warm period appears to be stable, ice ages are decidedly unstable. They show an occasional sudden warming, forcing the ice sheet to retreat, followed by an equally sudden cooling.

A cycle might complete inside just a few decades.

It has been suggested that these 'flips' may be due to an occasional cessation of thermohaline circulation. This would cause sudden cooling of the northern hemisphere, precipitating an equally sudden warming in southern latitudes. The global freeze would see-saw between north and south.

25 Certainty might be said to be the stuff of mathematics, often confused with science.

Climate change can clearly produce dramatic effects. In one incident, about 19,000 years ago, sea level may have risen *ten metres*.

The paleoclimatic record shows two things clearly: first, that variations of surface temperature and atmospheric carbon dioxide concentration correlate extremely well; second, that positive feedback is needed to explain the size of the changes in both. Small initial changes in surface temperature, brought about by a variation in solar irradiance, cause substantial change in atmospheric carbon dioxide concentration. This can then significantly change surface temperature, sea level, and glaciation.

Climate is evidently sensitive to small changes in circumstance.

Atmospheric carbon dioxide concentration has leaped upward by 50% since the start of the industrial revolution, and is rising rapidly. In and out of an ice age, it varied between 180 and 280ppmv, but is now up around 380, and is increasing by at least 1.5 each year. This corresponds to an annual addition of approximately 3.3gt to the atmospheric carbon reservoir.

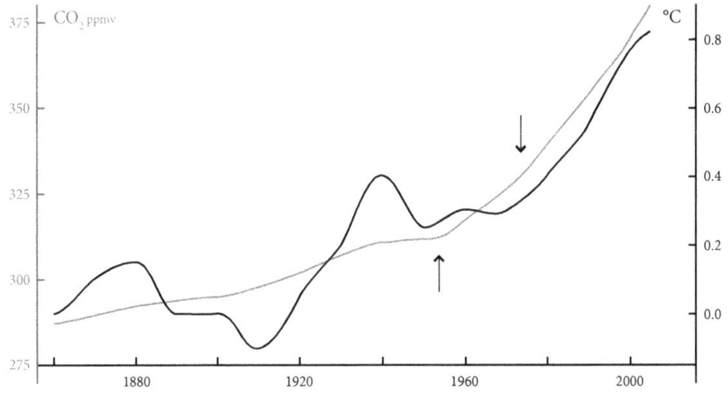

FIGURE 4.5 Temperature (from a base of 13.6°C) and CO_2 concentration over last 140 years.

Clearly, something dramatic, perhaps seminal, is occurring. There can now be no reasonable doubt that it is a result of our burning fossil fuel. The change just correlates too well with increasing industrial activity. A strong correlation can also be seen in the rise of the three primary greenhouse gases, aside from water vapour — carbon dioxide, methane, and nitrous oxide.

No one has offered any credible explanation of either correlation, other than as a consequence of the burning of huge amounts of fossil fuel.

Though we can predict the climatic effects in the short term (a few decades, perhaps), no one knows what they may be in the long term. We are dealing with a complex system where we neither know nor understand all the feedback, never mind the all interaction between feedback loops. Even if we were in possession of complete knowledge, predicting behaviour beyond a few decades is likely to be intrinsically impossible. In the words of one expert (to the extent that anyone can make such a claim), what we have done is to initiate a "remarkable experiment".

All we can do now is watch, and perhaps prepare for the worst.

One might hope that most of the excess carbon is a consequence of the early industrial days of burning coal. The evidence suggests otherwise.

There is no sign of any slowing as we switch to cleaner fuel.

Sadly, there is even a notable increase in the trend as of *circa* 1960. In the case of methane, one may even observe major economic changes, such as the decline of the Russian economy in the early 1990s. Furthermore, the temperature record *is* following the carbon dioxide trend, with a lag of a decade or two.

Two kinks are evident in the increase in carbon dioxide concentration (indicated by arrows in Figure 4.5), marking an acceleration. The rise is now understood to be exponential, not linear, and thus even less predictable.

Many systems, both natural and artificial, respond to a disturbance by following a sigmoidal (S-shaped) curve. Initially, we see an exponential rise (or decline), followed by a similarly shaped tapering off to some new level. With global warming, no one is clear when the latter might begin, or to what final temperature.

Ordinary weather forecasting is tough enough.

Two thirds of the evident *enhanced* greenhouse effect may be attributed to carbon dioxide emission, and about a quarter to methane, which is the dominant constituent in natural gas. No alternative explanation has withstood scrutiny. Modelling the climate is now popular internationally and very well funded. Nations compete vigorously to host the most powerful computer applied to this end.

No model has succeeded in reproducing the rise in temperature thus far seen without including anthropogenic emission.

The average surface temperature on our planet rose by 0.6°C during the 20th Century. Computer models augur an additional increase over the 21st Century of somewhere between 1.4 and 5.8°C. The IPCC place their current best bet on 2.5°C. This comprises 1.2°C as a direct greenhouse effect plus 1.3°C as a consequence of feedback. Some feedback effects are already in evidence, such as the rapid melting of glaciers and the arctic ice-cap. The latter is now receding at the rate of about 9% each decade.

These numbers do not sound large or frightening in themselves. However, it must be remembered that they are *averages* and refer solely to the Earth's surface. A little thought regarding the energy involved is perhaps more sobering.

It is now only a matter of two or three decades before the pre-industrial (but post ice age) atmospheric carbon dioxide concentration is doubled. This reduces the rate at which the surface can radiate heat into space by about 2%. Again, it doesn't sound much, but we are dealing with quantities of energy that make our paltry consumption pall into insignificance. Recall that, globally, we consume around 12tW. That 2% amounts to around 3,600tW — *three hundred times* what we consume worldwide.

To put it another way, we are trapping the energy liberated by detonating a standard 10mt thermonuclear warhead every twelve seconds.

One might hope for mitigation. After all, some of the feedback known is *negative*. For example, on land, plants are fertilized by an increase in carbon dioxide, and so digest more of it. Some measure of regulation arises this way. Unfortunately, there has been extensive tropical deforestation to add to that which has already taken place in the northern hemisphere (where the greater part of the land mass is found). If trees are not allowed to regrow then the short carbon cycle is broken. Although some re-establishment of northern forest is now being fostered, overall, something like an additional annual gigatonne of carbon is dumped in the atmospheric reservoir.

Our familiarity with carbonated drinks might lead us to think that the ocean would absorb some of the excess. Again, indeed they do. Our species actually liberates around 7.5 gigatonnes of carbon each year — 6gt from burning fossil fuels and 1.5gt from changes in land use and deforestation. Of this, around 45% (the 3.3gt mentioned earlier) remains in the air. The rest ends up in the sea. However, the sea cannot do better than this because only the surface layer (the first few tens of metres) can be effective in the short term. Mixing of these waters with lower layers takes centuries, or even millennia for the great ocean deep.

Fertilization of oceanic life may constitute another possible negative feedback. There is some evidence from the ice-core paleoclimatic record that the plankton population did increase at the end of the last ice age, as both temperature and carbon dioxide levels rose. This would have pumped more carbon out of the atmosphere and into the ocean. However, there is no direct evidence yet reported of that effect.

While no other significant negative feedback is immediately apparent, no fewer than three further positive feedback effects are predicted. Higher temperatures affect the respiration of land life, especially soil microbes, increasing their carbon emission. The stress of climate change is expected to reduce forest growth. Finally, as warming proceeds, there comes a point where methane is released, first from wetland areas, and second, and of greater long-term concern, from huge reservoirs of methyl hydrate in deep ocean sediments.

It seems pretty clear that we as yet have no reason to expect salvation from nature, and every reason to expect feedback to exacerbate climate change. As we learn more, our uncertainty actually *grows.*

Perhaps the most disturbing news is that the change in atmospheric carbon dioxide concentration is accelerating. Over fifty years, the average rise is just 1.3ppmv, that over recent decades is 1.6, but the average over the first five years of the new century lies in excess of 2.1.

Furthermore, some of the worst predictions are coming true decades before even the alarmists expected. The perennial arctic sea ice is melting. A full 14% was lost in 2005 alone. Within as little as a decade, it may become possible to sail to the North Pole. An open sea reflects far less solar radiation than an ice sheet — positive feedback of the first order.

Siberian permafrost is melting, liberating vast amounts of methane.

Even with our new-found computer technology, building a model that accounts for all known feedback is extremely difficult. One experiment, led by someone at Oxford University, managed around eight *millennia* of computation by encouraging ordinary people to download applicable software via the internet. (This is sometimes called 'grid' computing.) One outcome is an upper and lower bound for the Earth's surface temperature at the end of the 21st Century. While their lower estimate concurs with previous ones, the higher one predicts a rise of up to 11°C.

No one really knows the full consequences of a rise that large.

One we can be sure of is a lot more wind and rain.

Weather outlook: windy and wet

The British are famous for complaining about the weather. This is embarrassing since we don't have a lot to complain about. We just don't get out much. (It's often raining, or so gloomy you don't bother.)

Despite admirable attempts to make it interesting, to blokes at any rate, the weather forecast is pretty boring, though I feel compelled to say the most boring I've ever sat through was in Australia: "Gudday folks. Tomorrow is gonna be sunny and hot" — rather like today, yesterday, and the day before that. The weather may explain why so many poms have moved there. But still they complain. Australians complain about poms complaining.

Brits experience neither tornado nor hurricane. People who do seem strangely to complain rather less. Maybe they are just glad to have survived. Sadly, the record strongly suggests there are now more and stronger storms. Just a few weeks before penning this, assorted Caribbean islands and Florida were hit by three hurricanes in just four weeks — two Category 4 (Charley and Frances) and one Category 5 (Ivan), on the Saffir-Simpson scale (Table 4.2). Wind speeds of 240kph (150mph) were sustained.

Category	Wind speed *kph*	Storm surge *metres*	Central pressure *millibars*	Typical damage
1	119–153	1.2–1.5	980	To trees and mobile homes.
2	154–177	1.8–2.4	965–979	Preceding flood. Small trees down.
3	178–209	2.7–3.7	945–964	To buildings. Large trees down.
4	210–249	4.0–5.5	920–944	Evacuation up to 10km inland.
5	≥ 250	≥ 5.5	< 920	Buildings destroyed.

TABLE 4.2 The Saffir-Simpson scale of hurricane intensity.

Ivan was the strongest hurricane to hit Jamaica in sixty years.

Perhaps this was all within normal statistical variation, though insurance companies and climatologists are less than convinced. Insurance losses through US windstorm catastrophe rose tenfold from $5bn in the 1960s to around $50bn in the 1980s. Part of this rise is undoubtedly due to a population that is both increasing, over the period, and moving into more vulnerable areas. Part of it is due to an increase in the number and severity of storms. A similar story can be told in Europe.

Hurricanes form over tropical seas. Moisture evaporates, rising to form clouds. Condensation and a high altitude wind cools the air, which duly falls, to be warmed by the sea once more, creating a cycle. The Earth's rotation causes horizontal circulation, and propels the whole structure to higher latitudes.

The intensity of a hurricane rises with sea surface temperature and the depth of warm water. We know that the ocean surface has warmed by at least a half degree centigrade over just the last two decades, so it is hardly surprising that Category 4 and 5 hurricanes have become nearly twice as common as they were thirty years ago, or that the total power of all tropical cyclones has doubled.

2005 saw the record broken with *fifteen* Atlantic hurricanes, of which four were rated Category 5. One of these — Katrina — devastated New Orleans, leaving 1,800 dead, 500,000 homeless, and around $80bn in damage.

Whether or not a particular hurricane, or even a season of hurricanes, is attributable to global warming can always be called into question. The fact remains that we have seen *precisely* what we would expect to see, given what we understand of storm formation and what we know has changed in their nursery. Theory and observation are of one accord, thus far.

As the oceans continue to warm, storms will grow stronger and more frequent still.

As evidence for climate change accumulates, more significant effects may now be expected before the end of this century.

First, the pattern of rainfall, by both season and region, may alter. There is evidence that deforestation also produces this effect directly. It is probably impossible to distinguish between the two causes. Regardless, changing rainfall can be catastrophic, especially to people utterly dependent upon local agricultural production.

Two thirds of all water resources are used for agriculture and around a quarter for industry. Demand for water has been increasing exponentially since the end of World War II. Many nations share access to rivers, lakes, and underground aquifers. It has been said that water may be an even greater source of international tension than oil and gas. A similar lack of public awareness of the difficulty is apparent. My father was a water engineer. His profession struggled to get adequate investment against the perception that potable water was 'free' and plentiful everywhere. Supply in Britain is still precarious in some areas. Climate change will, to employ the political vernacular, "bring new challenges", even for the rich world.

Much of the poor world relies on predictable rainfall (or snowfall), though it may be more adaptable simply because its people are more used to calamity. Through lack of investment in water management, the poor world typically suffers a perpetual cycle of flood and drought.

Global warming is expected to increase rainfall overall, but at different times and in different places. In 2006, the UK Meteorological Office's Hadley Centre for Climate Prediction and Research predicted a dramatic increase in the proportion of the planet suffering drought by the end of the 21st Century. The figure for moderate drought rises from a quarter to a full half of the Earth's surface. That for severe drought rises from 8% to 40%, and that for 'extreme' drought goes from a mere 3% to 30%. Even these horrific statistics may form an underestimate because no account was taken of certain feedback effects, which are believed significant. At least a *third* of the world's land area may cease to support life within a century.

While the rich world can trade manufactured goods, financial services, and intellectual property, for food; the poor one cannot. Subsistence farming remains the norm. Lack of transport infrastructure limits distribution. Drought and famine will take their toll, and with no possibility of recompense. Poor people cannot afford good lawyers. (Even the rich have a tough time with that.) Legal redress is rather unlikely.

Agricultural productivity is expected to shift to higher latitudes, and possibly decrease overall, as a direct consequence of warming. The American "grain-belt" may well move north. As mentioned earlier, there may also be winners. Canada, Scandinavia, Russia, and Japan, may benefit in the northern hemisphere. In addition to finding vast tracts of what is now wasteland becoming productive, Russia is gaining access to enormous untapped fisheries as the ice retreats.

Meanwhile, in the southern hemisphere — home to the poor — there is much less land toward the pole. A turn of the globe shows perhaps Chile and Argentina gaining, but few others, and no regions of any great size.

A detailed study by the Oxford University Environmental Change Unit predicted a slight benefit for the rich world but a much larger cost to the poor one. Some form of redress would seem appropriate since global warming is largely caused by rich industrialized nations.

There are only three possibilities for the outcome of the massive desertification to come. Hundreds of millions of people can migrate to an area that remains productive, perhaps a thousand miles away. They might instead become entirely dependent on charitable aid, at immense expense, sustained indefinitely. Or they might simply curl up and die *en masse* — man, woman, and child — along with their entire culture.

Such is the ancient custom, when faced with *natural* catastrophe. But this disaster is brought, not by nature, but by the excesses of the rich world.

Living in England, with around two hundred and forty people per square kilometre, it is hard to believe, but nearly a third of the land on our planet is still covered by forest. Rainfall is influenced heavily by forestation and, of course, it is an essential element in the short carbon cycle, replenishing oxygen in exchange for carbon dioxide.

Trees are known to be sensitive to temperature. They cannot get up and move when the climate changes. In any given region, existing arboreal species will degenerate and be replaced by new ones, better adapted. With the expected rate of warming, there will be no time for such a response. Centuries, rather than decades, would be needed. While a greater concentration of atmospheric carbon dioxide fertilizes the growth of trees, the inability to adapt to changing climate will be the dominant effect.

Studies estimate *two thirds* of the world's forests are threatened.

Forests are also ecosystems, carrying a large proportion of existing species. Their failure would have a catastrophic consequence for biodiversity. Global warming is likely to further exacerbate the mass extinction already in progress, caused by the rise in human population, loss of habitat, and 'intensive' (industrialized) agriculture.

A fourth effect of ongoing climate change is upon sea level, which has already risen between ten and twenty centimetres (four and eight inches) over the last century. Experts expect it to rise a further ten to *ninety* centimetres (four to thirty-five inches) by 2100.

The IPCC believes a thirty centimetre rise most probable. That would be disastrous for the poor world, especially when added to other effects of climate change, such as more frequent and severe storms, and changing rainfall. Over a hundred million people live within a metre of sea level. Much fertile land would be lost in many parts of the southern hemisphere.

Again, the poor world suffers while parts of the rich one profit.

A one-metre sea-level rise is nothing compared with the consequences should the ice shelves above Greenland or Antarctica melt. Ancient coral heads, dating from before the onset of the last ice age, are found *six* metres above current sea level. These would have found themselves just below the sea surface when they were alive.

All it takes to melt the Greenland ice shelf, and reproduce that sea-level rise is a sustained temperature increase of three to five degrees centigrade. That much is probably inevitable now, before the end of the century.

This time the rich world suffers as well.

It would leave a third of London under water.

Even the possibility of a six-metre rise understates the full risk here. *Half* the world's water is locked up in these great overland ice sheets. (Melting sea ice has no more effect on sea level than a melting ice-cube does in your gin and tonic.) A melting of the west antarctic ice sheet (WAIS) would raise sea level an additional five metres, and that of the east antarctic ice sheet (EAIS) about an extra *seventy* metres.

It is hard to imagine civilization withstanding even the former.

I should immediately make clear that the prospect of either one melting is thought remote by most experts. To melt the WAIS is thought to require a surface temperature rise of 10°C and take thousands of years. The EAIS ought to withstand a 20°C rise.

But then the same people thought the Greenland ice sheet secure. Not only is it disappearing faster than anyone believed possible, but the WAIS is also showing signs of following suit. It had been thought that it would take several centuries for the Greenland ice shelf to melt. Now it looks like it might need just one, even with just a modest warming.

Science is playing catch-up with nature.

Last but by no means least is the effect of warming on human health. We humans can no longer migrate, as we once did. In the end, there are just too many of us. The political consequences are unthinkable.

France has recently suffered record heatwaves. As I write, five of the hottest years on record have occurred in the last seven. Many elderly folk failed to cope and succumbed to an early death. Those who study such things report that the increase in mortality due to excessive heat greatly outnumber the reduction due to fewer cold snaps.

Of still greater importance is the effect upon the war against the microbe. Bacteria, other parasites, and their insect carriers, all thrive in warmer and wetter climes. Malaria is still unbeaten, causing two million deaths each year. Around half the global population is currently at risk. That proportion will rise significantly with global warming.

The malarial map of the world will have to be redrawn.

Few, if any, of the many media articles on global warming and climate change talk about where all this might end. Partly, this is because nobody knows, but there is ancient precedent for a continuing rise in the concentration of atmospheric greenhouse gases. Recent research into the causes of 'prehistoric'[26] mass extinctions, including the demise of the oddly beloved dinosaurs, has turned up another explanation — one that appears to account for several such events. It runs like this.

First, volcanoes erupt, completing the long carbon cycle by emitting gigantic quantities of carbon dioxide and methane. Global warming, due to the greenhouse effect, follows. As the oceans warm, they absorb less and less oxygen from the air above them. Anaerobic bacteria, normally living in the deep, now rise to the surface to release immense amounts of hydrogen sulphide — the stuff that makes rotten eggs smell bad.

---

26 I have often wondered quite how something can precede history.

The reason we are so sensitive to this aroma is that hydrogen sulphide is even more toxic than carbon monoxide. It is so poisonous that much of Earth's land-life would likely be eradicated. Only creatures far removed from the sea might escape. But there is worse to come.

Hydrogen sulphide is lighter than air, and would rise, eventually making contact with the ozonosphere — a thin layer of ozone that filters out ultraviolet light from the Sun. Chemical reactions would then destroy the ozonosphere, leaving all life on land exposed to lethal radiation. It is this indirect effect of global warming that some now believe has been responsible for *several* mass extinctions. Paleoclimatology confirms extreme levels of atmospheric carbon dioxide immediately before each devastation.

Thankfully, the rise in greenhouse gas concentration caused by our industry and cars has not brought us anywhere near the one part per thousand that would appear from the record to threaten "oceanic anoxia". On the other hand, if the present rate of increase — 2–3ppmv per year — continues unabated then our descendents may well suffer that fate within ten generations. But this accounts for neither an acceleration in emission, resulting from global economic growth (currently 8% *per annum* in China) with 'dirty' generating technology, nor the real possibility of positive feedback.

Very slowly, we are waking up to the fact that we live on thin ice.

Environmental lobbyists and the so-called "green movement" are easily accused of "scare mongering" by publicizing the above projections. Pestilence and disease, sea-level rise, mass extinction, deforestation, famine, flood, and drought, are the stuff of the sensation-peddling mass media.

Real environmentalists have long realized that fear is not a great persuader. But fear is justified in the face of significant change coupled with great ignorance. A short circuit has been applied to the long carbon cycle. Ecologically significant quantities of greenhouse gas have been released. Surface temperature is rising, and the climate is already changing in response. Feedback will take effect, and either restore normality or bring catastrophe. All we can do is watch and wait.

And maybe pray.

A lot of hot air

It should never be forgotten that the true climatic consequences of doubling atmospheric carbon dioxide are actually *unknown*. All predictions discussed above have been made as a result of expertly constructed computer models. We would be right to harbour suspicions of these, not least because they are vulnerable to any personal or professional bias their masters may sport. But it would be equally wrong to dismiss them. We have no other guidance. To ignore the council of experts, in consensus, is the mark of a fool.

Models form a spectrum from "business as usual" to scenarios that assume widespread adoption of ecologically driven policies. Even at these extremes, there is massive opportunity for variation. Economic activity, population growth, war, and even changes in fashion, can have significant effect. The coming degree of technological innovation is undoubtedly critical.

Two organizations that construct models and investigate scenarios are the World Energy Council (WEC) and the IPCC. Their conclusions are broadly consistent, particularly with regard to "business as usual". Under the assumption that no ecologically driven policy will emerge, both predict the doubling of atmospheric carbon dioxide concentration by around 2060, with increase continuing well into the next century unabated. The only model that predicts stabilization at any point this century is the WEC "Scenario C" which is *fully* ecologically driven. Here, the concentration ceases to increase around 2070 and levels off at about 400ppmv.

The WEC made moderate assumptions globally for both population and economic growth. Population is assumed to reach ten billion by 2050 and twelve by 2100 with average economic growth stable at 3.3%. Energy intensity (energy consumed per dollar product) has been falling by about 1% each year for decades, because of improvements in energy efficiency and change in the very nature of products. Scenario C assumes an annual fall two and a half times greater. Energy demand would remain high, but half the primary supply would be generated by sustainable means by 2100. Technology transfer from rich world to poor would be utterly uninhibited.

Carbon emission would fall to a level nearly 60% below that of 1990.

No radically new technology is required, nor is there a case for any increase in the centralization of power generation. Nuclear power, for example, is *not* an 'essential' substitute for burning fossil fuel. Scenario C assumes only 11% of energy demand met in this manner, compared with less than 7% in 2000. In fact, the energy industry might become much more communal, and also much more appropriate for the developing world.

A large number of small power stations would be more flexible, taking advantage of locally available sustainable resources. They can also eliminate the expense and inefficiency of long-distance transmission.

Investment would be distributed, instead of concentrated, and thus open to smaller organizations — even to the village and individual.

We could all regain control over something we each depend upon.

As we have said, fear of climatic catastrophe ought not to be used to persuade an electorate of the need for action. Caution and reason should suffice.

There is another fear to be shunned — the fear of *regress* — a dread that we should somehow slide 'backward' toward the darkness of the cave.

As a scholar, raised and tutored in objectivity, I struggle to understand how we collectively cling to the notion of 'progress', and revel in the most sophisticated technology, however pointless its application, and however far removed from popular understanding. Perhaps it brings us hope.

It is not technology that will keep us in the light, but *science.*

We may be sure of two things. First, there really *is* a problem. Second, while our routine may change, a high quality of life *can* be maintained.

To know a little is always a dangerous thing, but as long as our commitment to science remains as strong as it is, we shall learn enough to survive our own success. No one is going back to the cave. Quite the reverse. Many still in darkness could be brought into the light. What we arguably need is not higher technology, but better and more appropriate technology, which might liberate the poor world, while saving the rich one.

If only the power of concentrated capital would allow.

Traditionally, governments do not respond to any event that has yet to occur, ...or so they may say. In fact, governments do indeed endorse what has become known as the "precautionary principle". They *frequently* act to counter something before it has happened.

One example is the maintenance of a standing army.

For some reason, precautionary policy is less than popular when it acts against possible environmental change. It would seem to depend whether they are convinced that the change in question may indeed occur. If you truly believe an enemy may otherwise invade then you build your defences. If you truly believe the ozonosphere may be destroyed then you negotiate a treaty (in Montreal) to ban the chemicals whose release would bring that about.

Enough governments became convinced by 1992 of the possibility of significant climate change to meet up (this time in Rio de Janeiro) and agree a (precautionary) response. They set up the United Nations Framework Convention on Climate Change (UNFCCC) which acknowledged that the rich world was primarily responsible, and must drastically reduce its carbon emission. Meanwhile the poor world would be allowed to increase its contribution as a result of economic and population growth.

A stated objective was to stabilize emission at a level that would prevent "dangerous anthropogenic interference with the climate system". Lack of "scientific certainty should not be used as a reason for postponement of measures" — the precautionary principle at work. Signatories were committed to report national inventories of carbon sources and sinks, publish measures to reduce emission, develop and transfer applicable technology for emission reduction, protect and conserve all sinks and reservoirs, and communicate policies with the aim of reducing emission to 1990 levels.

America — far the largest emitter — both signed and ratified, along with a hundred and eighty-four other nations.

The convention came into effect in 1994.

On its own, the UNFCCC set no protocol governing who would do what and by when. It was only ever meant as a starting point. This is why the parties met again, this time in Kyoto, to negotiate the "Kyoto Protocol".

Everyone subsequently had their own target to meet, to reduce their emission to a certain level with respect to what it was in 1990. Britain was to reduce it by 8%, America by 7%, and Germany by 25%, for example. Russia was merely to stabilize hers, while Iceland could actually increase emission by 10%. Judgement regarding what is and is not acceptable — for example, a contribution by forest expansion — is referred to the IPCC.

One controversial measure was to allow *trading* in emission.

The Kyoto Protocol would not become binding unless at least fifty-five nations signed and ratified, accounting for at least 55% of global emission.

Notably, the American government under Bill Clinton signed but were later ousted by George W. Bush, who refused to allow ratification. Given that America contributes a full quarter of global emission, this was less than constructive. However, Russia signed and on the 22nd October 2004 ratified. Even without America, more than 55% of global emission was then accounted for. The protocol could come into force.

Kyoto promised a 12.5% reduction in global greenhouse gas emissions by 2011, yet the only model that yields stabilization of the greenhouse effect (WEC Scenario C) requires a reduction of 24% by 2050 and 60% by 2100.

Some respected experts are beginning to revise their recommendations to be considerably more demanding. A report published in 2006 by the UK Tyndall Centre for Climate Change Research, which brings together six highly regarded university departments, calls for an emission reduction of *70% by 2030* in order to stabilize warming at just 2°C — the accepted figure beyond which serious climate change will ensue. This had implied limiting atmospheric carbon dioxide concentration to 550ppmv. Improved understanding has now established a revised limit of just 450ppmv, to give a "reasonable to high" probability of avoiding catastrophe.

All trends — in emission, in its consequences, and in what is required for mitigation — appear to be going the wrong way.

Meanwhile, progress toward an effective and globally inclusive international protocol is far from encouraging. The 2009 Copenhagen meeting ended in what can only be called farce, with no meaningful agreement.

So what?

With just 4% of global population, America contributes a quarter of global carbon dioxide emission. Of this, a full third is from transport, and more than a quarter from road transport. While American industry is getting cleaner, largely through a move away from coal, transport emission is rising, due to increasing mileage and decreasing efficiency (the taste for heavier cars). Overall, carbon emission rose by 16% between 1990 and 2002. 38% of that growth is attributable to transport, *i.e.* trucks and private cars.

The US Federal Government has finally accepted that scientific consensus has indeed been reached, but currently lacks any policy that would significantly reduce emission. As yet there is little public awareness, never mind acceptance, of the need for change.

In Britain, the Labour Party came to power in 1997 with a pledge to reduce greenhouse gas emission to 20% below 1990 levels by 2010, trumping a recommendation by the Royal Commission on Environmental Protection (RCEP) to do so by 2020. A full 18% cut had been accomplished as early as 2007, with a steady annual reduction of about 2% each year. However, none of this owes anything to transport, whose emission had *risen* by 9.5%.[27]

Air travel grew massively in the same period, with emission rising 53%, although this accounts for only about 1% of the total. Road transport, on the other hand, grew by just 9%, but from a very high level. Vehicle energy efficiency improved a little (7%), though the taste seems to be catching on in Britain for ever heavier cars. Growth in mileage led to the emission share of road transport rising from 22 to 26%. At the same time, rail travel rose in popularity by more than a quarter and simultaneously reduced fuel consumption by almost half. Rail now provides about 6.5% of passenger travel with only around 1% of the carbon emission.

We noted in Chapter 2 that car passenger mileage is overstated because proper account is not taken of how often the driver is merely a chauffeur. Even then, the average recorded vehicle emission is over 190g/pass-km[28] — more than three times that of a train. Yet again, the bus fails to resolve the issue with an emission almost as great as the car, by the same measure.[29]

One way this appalling statistic can be mitigated is by actually applying the law of the road. A 2007 report to the UK Parliament noted that enforcing the current 70mph (112kph) speed limit on motorways would bring a reduction in vehicle emission of a full 10%. This is rendered credible by the French accomplishment of a 19% reduction through a crackdown on speed.

Meanwhile, two nations with vastly greater populations — China and India — are industrializing, and following the dreadful example they have been set. As well as building cheap and dirty coal-fired power stations, they are trading in their bicycles for cars. If the rest of the world becomes as profligate as the US, there will be a *tenfold increase* in carbon emission.

There would be no hope of controlling sea level or climate change.

27 The fall in emission is largely accounted for by the end of heavy industry in Britain, halving industrial emission over the final decade of the last millenium.

28 Implying 53g carbon emitted per passenger-kilometre. Care is needed to distinguish a measure of the element (carbon) from that of the compound (carbon dioxide).

29 Recall that we assume a generous occupancy of 50% with any shared transport.

4.3 In the vicinity

Divided we fall

Most people have more immediate problems than the global environment. Although remote affairs have a habit of turning up unexpected and biting harder, those nearby interfere with daily life more often. We care mostly about where, how, and with whom, we live and work, and shut out the grander threats.

I am lucky enough to live in an ancient English village. Its streets are narrow, and its buildings of a very human dimension. A thriving town before the nearby city of Oxford was borne, it still has many amenities to bring its folk together — a church, a school, a shop, two pubs, a playground, two bus stops, and a railway station.

These are where we met our neighbours, when we arrived. While the shop provides an excellent exchange for news, the train works best for introduction and conversation. A community emerges from the sharing of such things.

You have to grow up in a village, or a small town, to appreciate what it means to *belong* somewhere, or the reciprocal: how a place can belong to you. In times past, this would never have been a matter for conversation. It would simply be accepted, without question, that the place where you live would form an indispensable part of what you meant by the word 'home'.

Movies used to be made about such things, like Frank Capra's immortal *It's a Wonderful Life*.[30]

As the makers of those movies sensed, after the war, the very notion of community was under threat. Many people have now grown up without ever knowing it. As a result, they will never be able to understand it. What they have never known, they will never miss. With interdependent liveli-

30 Adapted from the book *The Greatest Gift* by Philip Van Doren Stern.

hoods and shared amenities, the village and small town stood the test of time. They were woven from what we are.

Of course, the city too knew community, and has a history almost as long as the village. Just as the village provided a local *focus*, there always had to be one for regional authority and trade. As the city grew, it would typically compartmentalize into distinct locations. They would take names that would hark back to their roots, like Greenwich Village or Camden Town. The fabric of community would emerge.

Many cities overran outlying villages, which had a curious habit of retaining their original focus and identity. London is a classic example. With each new transport innovation, more settlements would be first connected to, and then absorbed by, the expanding metropolis. But they refused to dissolve. Instead, the city itself became an aggregation of interconnected foci, gaining interest and diversity.

No planner has yet succeeded in deliberately replicating the effect.

Sprawl is a new kind of environment that knows no life save that bound within a vehicle. Cycling is dangerous. You may well find little provision for walking — a narrow path perhaps with few, if any, crossings. Motor traffic will, in any case, make walking difficult and unpleasant.

An attempt to perambulate may even be met with abuse.

Your field of view comprises a wide expanse of tarmac, a sea of vehicles, parked or attempting progress, and a plethora of road signs. Trees and shrubbery are removed as obstructions. Grass wastes space on which another vehicle could stand.

This new form of habitat is also *ugly* in the extreme.

Back in the land BC (before children), my partner and I played sailboats in the Mediterranean and Aegean. Some day, we will train up our boys and let them take the helm. That part of the world is littered with the remains of a civilizations which shared a strange idea; they put art on the *outside*, instead of hiding it all in a museum.

The beauty of Graeco-Roman architecture is breathtaking and it was always open. Statues would line a market-place for all to see. True, they enjoyed a better climate, but even where we have the opportunity to celebrate space and environment, it is usually wasted.

Nature is also typically altogether excluded.

The ancients could assume you would arrive on foot, or perhaps hoof. This gave them an immense advantage in available space. Our new and alien landscape commits a fatal crime at the outset. It gives over most space to the new means of arrival. The automobile (car and bus) requires more territory than any other mode of transport. Every open area disappears under concrete and tarmac. Buildings are lost behind road signs. Pedestrians become confined, obstructed, and threatened.

A two-lane road consumes three times the land occupied by a cycle-way, while offering far lower capacity.[31] Car parks become so large that buses have to be used to ferry people to the buildings they serve. Road space increases to occupy up to half of the land devoted to any town.

Despite this, parking becomes hard to accommodate and thus very expensive. In London, charges run well over £5 an hour. Even then, parking must often make way for other commercial or residential purposes, worth even more. While the cost of underground parking is ostensibly lower, capacity is limited and crime a greater threat. High-rise car parks win few awards for architecture. Again, they attract crime and make users feel insecure against it. But the primary drawback of inner city parking is the proportion of the estate given over to causeways converging upon it, and their dominance over the space through which they pass.

Motor traffic hardly makes for a pleasant place in which to shop or while away an hour in a street café. A desirable pedestrian environment is also one in which you can move about freely. Every road presents an intimidating barrier to the walker. The busier the road, the worse the obstacle. Modern pedestrian crossings, unlike their predecessors, always give priority to the vehicle, even though it rarely carries more than one passenger.

In Britain, there is a long-cherished tradition of providing for the pedestrian. We have pavements. It is illegal to drive on them, though, as we noted in Chapter 1, parking on pavements has become common practice. Many areas have had their pavement confiscated in this way. One sees the elderly and prams regularly forced into the road. Despite literally a ton of evidence, the local constabulary remain utterly disinterested.

The car has undoubtedly intruded upon, and damaged, urban life.

31 Especially when pedestrian crossings are taken into account.

Even when drivers park in the road, it adds considerably to pedestrian danger and nuisance (see Figure 2.5). It can also restrict flow. Parking restrictions are increased yearly. More drivers chase fewer opportunities. As they are forced back into the suburbs, they encounter yet more frustration.

Enter the exclusive "residents' parking zone" or RPZ.

For many years, I taught computer programming. At its heart, this is about *algorithm*. The term harks back to the time when Baghdad was the world's centre of learning, which reached barbaric Europe slowly via the Moors. We write text in Latin script but numbers in Arabic because we learned mathematics from Arabic books. One of the most famous and influential was by a man called Al-Khwarizmi who wrote about arithmetic using 'modern' Indian decimal place-value notation. His name lives on as 'algorithm', which refers simply to method, or procedure.

Some algorithms are more elegant or powerful than others. A *recursive* algorithm is one that refers back to itself at a more specific level — a little like the way your reflection can itself reflect, in ever decreasing scale, in a pair of opposing mirrors, as you sometimes find in a lift, for example.

Cats are remarkable animals. They have perfected a recursive algorithm to conquer territory and take over the world. Those who keep them are in little doubt of their success. Somewhere there must be a university for pussies, spreading the word. Their method is very simple, like all the greatest algorithms. All they have to do is defecate in the garden of someone who does *not* keep a cat. The victim then faces a choice between living in the local feline toilet and acquiring a cat of their own. Voilà! — a home for another moggie who can join in the fun and increase the rate of conquest.

RPZs work the same way. All you have to do is to park somewhere that has not acquired that status. Voilà! — another one is borne.

People really don't like strangers parking outside their house. Once it starts, they demand that their area becomes exclusive. As a result, drivers park in streets a little further out, and the process recurs.

Direct regulation, and punitive taxation, is also increasing. Local authorities are starting to levy a charge on company parking spaces and even on entry to town or city centre. The London Congestion Charge has been effective in reducing traffic but has been criticized for raising too little money for improvements to shared transportation. These were always intended to follow later. Pain was always part of the recipe.

Some believe that things were the wrong way around. Better alternatives should have lured folk *voluntarily* away from their car.

The result of all this is that, while you can still venture forth in your car, and might perhaps penetrate the congested conduit, you may then find it unaffordable or impossible to park when you arrive.

With no rail or bus service on a Sunday, we once set out in our car to visit a museum and restaurant in town *en famille*. I had not realized how much things had changed. We gave up and came home again. Combining a lack of an alternative to the car *and* the chance to park is less than constructive. Everyone loses —the museum, the restaurant, and my family.

The utility of the car is in organic decline as a direct consequence of our over-reliance upon it. In many towns, capacity of neither road nor parking can now meet demand, which continues to grow rapidly. Disinvestment in alternatives leaves transport *per se* dysfunctional, with catastrophic consequences for both lifestyle and the local economy.

Real choice, beyond the automobile, is urgently needed. Alternatives will need to make far better use of available conduits, and impede neither pedestrian nor cyclist. Beyond these, ultralight trams and various forms of light electric vehicle (LEV) wait in the wings.

I dream of peaceful piazzas with a café culture.

And let's put art on the outside this time.[32]

You must be choking

One consequence of choosing to use a car is that you pollute your *own* environment. The worst place to be for air pollution is inside one. The second worst place to be is at the height of a child on the pavement.

Air pollution is an understated health problem. According to British Government figures, for every person killed in a road traffic accident (RTA), somewhere between four and eight (12–24,000pa) people die as a result of air pollution.[33] Victims tend to be those who already suffer respiratory ailment, but their lives are surely no less valuable. My own is among them.

---

32 I believe I am in the majority when I declare that, to be art, it *must* please the eye.

33 Department of Health, Committee on the Medical Effects of Air Pollutants.

They further estimate an additional annual cost, in health care and lost production, of around £20bn. The cost in ill health remains unknown.

In town, road vehicles are the dominant source of air pollution.

Air consists of 78% nitrogen and 21% oxygen. Gasoline and diesel are mostly hydrocarbon that needs little encouragement to burn in air, liberating oxides of hydrogen, carbon, and nitrogen. Since sulphur is a common contaminant, some oxides of sulphur also emerge, along with hydrogen sulphide. Thankfully, lead is no longer added to fuel as it was found to cause brain damage, notably in children.

Because an internal combustion engine (ICE) burns fuel at a modest temperature, it does not fully consume it. As a result, some larger molecules are also emitted, known as "volatile organic compounds" (VOCs). Evaporation and leakage, from the tank and its filling, also liberate VOCs — the carcinogen benzene, for example.

Carbonaceous particles also exit the tail-pipe, with VOCs embedded. Our airways trap the large ones, but those smaller than about 10μm — so called PM10s — are known to cause serious, sometimes fatal disease.

PM10s can travel miles from their point of emission and stay airborne for days or even weeks. They affect breathing, exacerbate asthma, and cause lung damage. People with existing heart or lung conditions can die as a result of exposure. Particles smaller than about 2.5μm travel deep into the lung where the VOCs they contain can cause cancer.

Filling stations are hazardous places. Not only are they subject to carcinogenic vapours but leakage also accumulates underground threatening the aquifer below. A hemisphere of pollution lies beneath every filling station, gradually expanding. It can even cross the street, putting its neighbours at risk. While it *can* be cleaned up, the cost is huge — another as yet uncounted external cost of motoring. Where such work has been done, the poor old taxpayer usually foots the bill, not the polluter.

Of the oxides that an ICE emits, that of hydrogen is just water vapour. Others are more of a problem. Carbon dioxide (CO_2) is usually harmless to breathe but poses the threat of global warming and climate change, as we have seen. Carbon monoxide (CO) is a potentially lethal toxin that hangs around for days before combining with oxygen to form more carbon dioxide. It reduces oxygen transport within our bodies, causing anything from dizziness and loss of concentration to death.

Of the oxides of nitrogen emitted, 10% is nitrogen dioxide (NO_2). 90% is nitric oxide (NO) but this quickly oxidizes to form more NO_2. (The two are sometimes referred to collectively as NO_X.) A much smaller amount of nitrous oxide is also emitted (N_2O) — another greenhouse gas, significantly more powerful than CO_2.

Nitrogen dioxide is known to exacerbate respiratory disease and may increase susceptibility to infection. On warm sunny days, photochemical reactions yield copious amounts of ozone (O_3). Above a certain threshold, ozone causes breathing difficulties in some people, and is known to induce attack in asthmatics.

While there is general recognition of certain effects of air pollution on health, the variation in degree of risk with level of exposure is unknown. It is not an easy thing to measure. You cannot experiment on humans, nor can you easily distinguish the cause, once someone has fallen sick.

Once again the "precautionary principle" might apply.

There may also be other effects, not yet recognized. For example, both the incidence (number of attacks) and the prevalence (number of sufferers) of childhood asthma has risen in a way that correlates well with the rise in air pollution, although it correlates with other things as well. Again, it is very difficult to establish a causal link. However, a link would be consistent with both the correlation *and* with the prior knowledge that some pollutants at least exacerbate the ailment.

Something has caused the doubling of asthma among children within a single decade.

It took many years before it was finally accepted that lead emission from road traffic caused brain damage. Establishing a causal connection is not about performing statistical trials. These can demonstrate only association. They are, in any case, subject to "systematic effects", whereby the outcome of a trial can be biased by an imbalance in the sample of the population chosen. When attempting to associate asthma with air pollution, you would have to account for smoking in the home, for example.

Given human lifestyles, there are just too many possible factors.

Until we know more about the disease, and the rôle of pollutants, we are left deciding whether or not the precautionary principle is applicable.

To experience an asthma attack is to experience drowning. At the tender age of eleven, after an evening on the beach, I remember believing I was dying. I simply could not breathe. I could not understand the lack of panic around me. My mother confided in me, shortly before she died, that she too believed I might not make it through that night. There was nothing she could do. Either I'd be there in the morning or I wouldn't.

Children get ill a lot. You can't call an ambulance every time.

Asthma is also debilitating. Although I grew out of regular onslaught in my teens, I remain susceptible. While in Australia, I wanted to go diving and sought the required medical certificate at the Royal Adelaide Hospital. Although I had suffered the condition for decades, British health care had mustered not a single test. In Adelaide, three simple inexpensive tests confirmed it was indeed asthma and established useful parameters. I could dive but only at significant risk.

Moist air, particularly moist salt air, triggers attack. It had done so in their lab as it had on that beach when I was a child. I missed out on diving on the Great Barrier Reef, something I had dearly wanted to do. But then I missed out on swimming throughout my childhood.

To inflict such limitations on a generation of children is abhorrent.

Over the last quarter century in Britain, childhood asthma has increased *sixfold*. Diagnosis of asthma among preschool children rose in the 1990s alone from 11% to 19%. This represents suffering and disability an a massive scale.

Some suspected causes declined in the period, and can be excluded.

My own asthma may have been the result of some combination of genetic predisposition (my maternal family were prone to tuberculosis), a draughty home in a damp climate, made worse by gas cooking and my father smoking around me.

Passive smoking and exposure to damp are on the wane, and so seem unlikely culprits. An increase in reporting the condition can also be ruled out because one would then expect to see more cases with less severe symptoms in the surgery. The number remains the same.

While other explanations remain possible, air pollution as a cause of respiratory disease, such as asthma, is quite consistent with the facts.

Incidence seems to vary with urbanization. America is far less urban than Britain and suffers less, though the rate has still doubled between 1980 and 1995, from 3.6% to 7.5%, in those under eighteen. In the city itself, things appear to be much worse. One expert recently claimed that 40% of children in Perth, Australia, suffer asthma by the age of six. They also blamed air pollution as the primary cause, though admittedly with more of an eye on household contaminants such as cleaning agents.

A recent $18m ten-year epidemiological study in California was more direct. Among children who played outdoor sports, it found three times as many developed asthma where the air was badly polluted by ozone. High ozone episodes correlated with school absence due to respiratory illness. The development of both the lung and its proper function was also found to be impaired by a cocktail of air pollutants typical of vehicle emission.

Another US study found that while the prevalence of asthma among Elementary School pupils nationally at the time was around 9%, that in schools close to a highway rose to 24%, and would vary with proximity.

The evidence would seem strong that air pollution not only precipitates and exacerbates asthma attacks, it causes the disease itself.

There can be no doubt that cars are a primary cause of air pollution. Figures for Britain in 1992 showed transport responsible for 90% of carbon monoxide, 60% of nitrogen dioxides, 48% of VOCs, and 48% of PM10s. Of this, cars contributed 88%, 72%, and 84%, respectively, while heavy goods vehicles (HGVs), including buses, contributed 84% of PM10s.

Quite apart from the cost in human suffering, the financial cost of health care provision is immense and rising rapidly. Respiratory ailments are not cheap to treat. The direct cost of treating asthma alone in Britain is around £1bn each year. Costs follow incidence and are estimated to have risen in the US from $6bn in 1990 to nearly $14bn in 2000 — yet another figure to add to the external cost of motoring.

It is now possible to reduce air pollution caused by the ICE.

The "catalytic converter" employs a catalyst, such as platinum, to promote chemical reactions that reduce VOCs and nitrogen oxides. Nitrous oxide (N_2O) emerges in place of nitrogen monoxide and dioxide. Carbon monoxide is also largely eliminated in favour of carbon dioxide. Greenhouse gas emission is arguably worse, while toxic emission is relieved. Since 1992, transport emission of carbon monoxide has fallen by more than a half and nitrogen oxides by over a third, despite a steep rise in traffic and mileage. This is primarily due to the widespread adoption of the 'cat', although there has also been some benefit from a shift to diesel from petrol.

Cats have little effect on particulates, whose emission from vehicles has fallen by about a third, thanks to tighter regulation and improved engine design. However, other sources have reduced their emission by more than half.

Unlike their feline namesake, cats have but one life, which may not be a long one. They are easily damaged and effective only if cared for. They are also *very* expensive to replace, constitute a fire hazard, and consume scant resources. If the fuel is sulphur contaminated, hydrogen sulphide will make its presence known with the stink of rotten eggs. Recall that H_2S is highly toxic.

Complete combustion of any pure hydrocarbon would yield only carbon dioxide and water vapour. Carbon monoxide forms when the burn is short of oxygen or at too high a temperature. Idling and deceleration can cause this. To make a cat work, the air/fuel mixture is controlled to be always "chemically correct", at a 'normal' 14.7 : 1.

Greater fuel economy and thus less emission can be achieved by varying the mixture so that a 'lean' mixture of up to 22 : 1 is used when full power is not required. High temperature "lean burn" ICEs were developed around the same time as the cat and have an advantage in not reducing power output. Sadly, you cannot use both at once since cat efficiency reduces rapidly as you move away from a normal mixture.

"Lean burn" has since been largely abandoned in favour of the cat.

Either way, we are left with greenhouse gases, and breathing particulates, even assuming every vehicle is fitted with a cat, and that it is working properly. Many vehicles have no effective means of emission reduction whatsoever. Few nations operate an effective vehicle emission testing and control programme.

On the day I write this, the local daily paper is running a headline "Choking for Change". A new report by the combined county and city councils has concluded that national limits on air pollution in Oxford are exceeded most of each day. Cars have largely already been evicted following adoption of the Oxford Transport Strategy (OTS) in 1999 (see Chapter 2).

The new polluter is the bus.

Air quality improved significantly following the introduction of OTS. There is some evidence from local hospital admissions that respiratory health also improved. Both are now under threat from the subsequent increase in the number of diesel buses and their emissions. In some parts of the town, nitrogen dioxide levels have more than doubled. The city council has found that nearly two thirds of NO_X pollution is now caused by buses.

Another report, privately commissioned the previous year, declared Oxford to suffer the worst air pollution in the country. The new report seems to confirm this.

Swapping cars for buses is evidently no solution to air pollution.

Up for the ride

We have seen how children have lost their independence, liberty, and health, through airborne pollution, with the rise of the private car. The effects on their health through loss of exercise, is only now emerging as yet another effect of the dominance of the automobile over our lifestyle.

A gobsmacking *third* of American adults are now clinically obese, up from a still woeful 13% in 1960. Nearly two thirds (*i.e.* the majority) are overweight. Europe is following suit. A full *quarter* of adult Brits are now obese, along with a horrifying *one in six* children. While the number of adults affected has doubled, the number of children has tripled.

Something called the "body mass index" (BMI) gives an admittedly coarse measure of how appropriate our weight is to our height. Using metric units (kilograms and metres), a BMI over twenty-five usually indicates we are overweight, and over thirty that we are obese. Some account should be taken for 'build'. (Your partner, like mine, may disagree!)

Obesity is not just something ill-advised. It is a serious medical condition that can threaten life through a range of diseases, especially of the heart and circulation.

Eating habits have become the standard target for blame. Fast food outlets have suffered accusations that they encourage bad habits. But it is difficult to reconcile this with what is known about metabolism. Our bodies have a built-in mechanism to eject excess fuel. Genes are then blamed for 'predisposing' some to suffer a disability in this respect. Drugs to trigger the mechanism have been developed. The fact is that it seemed to work fine in our ancestors, who were usually prone to hard physical labour. It still works fine in the typical athlete or construction site worker.

While, undoubtedly, some people do indeed suffer a disability, we should adopt the simplest explanation for the majority — that the mechanism of rejecting surplus carbohydrate is far less effective when the subject fails to get enough exercise. Indolence is the most likely cause of obesity.

A sedentary life might also be expected to take its toll on mental health. Exercise is well understood to improve a sense of well-being, and correlates with a lower incidence of depression. It may be no coincidence that the mental health of children has declined rapidly over the last fifty years, along with their physical health.

There can be no doubt that, in the rich world, children now get far less exercise than their grandparents did. Where once walking or cycling to school was the norm, children now arrive strapped in the back of a car. Where the bike was once the liberty of every child, it now lies disused in the garage. Cycling has simply become unsafe, especially for the young.

All the common land on which I once played football has now been 'developed'. Designated playing fields are all too often only accessible by car, or by crossing dangerously busy roads.

As diseases of the indolent take their toll around us, health may well enjoy greater influence over our personal choice in how to travel. We may well decide to walk or cycle more often. But health is not just our private concern. Those who fail to take proper care of their own health rain down costs on the rest of us. Whether medical provision is paid for by the State or private insurance, we all share responsibility.

While some might call for the irresponsible to be denied cover, it seems unlikely that a majority would support "lifestyle assessment" to influence either charges or service to a truly significant degree. Letting people die in the street is not an option in a civilized society. The State (*i.e.* we taxpayers) may very well be expected to "pick up the tab".

It will be a big one. Estimates for the increased annual cost of healthcare due to indolence in the USA are of the order of a *hundred billion dollars.*

Without doubt, there are other factors, but it defies reason to exonerate the car, especially with regard to children. It is the car that has replaced the walk to school, and denied them safe use of their bikes. It is not "stranger danger" but the busy road that has driven them indoors, and thus denied them normal outdoor play. The use of a car for every single journey has denied their independence and inured them to passivity and indolence.

Never has childhood been so impoverished.

See me, hear me, touch me

Living near a road, as the majority in the increasingly urban rich world do these days, means tolerating various forms of pollution. We have already noted the sheer ugliness that arises when an area becomes saturated with cars, filling the field of view, and turning front gardens into car parks.

Street lighting would now *appear* essential. In Britain, regulation *requires* all newly built residential and commercial avenues to be lit. No consideration is taken of any local objection.

Gas discharge lamps are the most economical. These emit light with a very narrow spectrum and a high intensity, which penetrates further than that of older filament bulbs. Much of it ends up heading skywards, with a significant proportion then scattered back down in all directions. This is sometimes called 'sky-glow' and is often orange — the colour of a sodium discharge. A starry night-sky is replaced by an eery blaze that suggests the work of the Dark Lord behind the gates of Mordor. A recent move to using a mixture of elements, including mercury, has resulted in whiter light, but with much the same effect. The beauty and wonder that has inspired poetry and science for thousands of years has been systematically annihilated.

Wide-area illumination confers no real security. It just creates the illusion of security. Glare can conceal rather than reveal, since it interferes with natural contrast, for which our eyesight has evolved. There is even some evidence to suggest that night illumination *assists* the criminal by helping identify opportunity, such as an open window.

After all, nobody likes to work in the dark.

Those up to no good also welcome the shadows in which they can hide. Sensor-triggered security lamps might sometimes attract attention. However, the majority of break-ins take place in broad daylight.

A Government website on home security openly states that harsh lighting is no deterrent to the criminal. Power failures, and even deliberate experiments where street lighting is turned off, show no corresponding increase in crime. In fact, they show the reverse — a small decrease.

The real reason we light up every street has to do with instinct. Our ancestors spent millions of years with real nocturnal dangers and a justifiable fear of the dark. Our sole security was the light of a fire.

Undue credit is also awarded street lighting with regard to road safety. I can find no direct evidence that it brings about any general improvement. As with security against crime, it merely carries that impression.

In one database of over 4,700 fatal road traffic accidents (RTAs), only two listed poor street lighting as the "precipitating factor", while 970 listed excessive speed — the second largest contributor, after loss of control. Yet safety is invariably cited as the primary justification for our massive expenditure on street lighting. The cost in light pollution is no more counted than that in cash, energy, or the resulting carbon emission.

Road safety *does* depend upon visibility, but it is an error to think that visibility automatically increases with illumination. Brighter lights do *not* imply clearer vision. In fact, they all too frequently yield the opposite. First, there is glare to consider. The presence of a bright light within our field of view interferes with our ability to interpret the scene. It distracts, may cause discomfort, and prevents the eye adapting to whatever light level is ambient. Second, it is *contrast* that counts.

As the Sun sets, we perceive only a small proportion of the true change in illumination. Our eyes are superb instruments and adapt to the huge changes in the amount of light that enters them — from brilliant sunshine to faint moonlight, or even starlight. Apart from the loss of colour vision, we remain unconscious of the true change in illumination.

The human eye is at least as sensitive as the finest electronic instrumentation we can make. I once used a photomultiplier capable of detecting just a few dozen photons per second. A young, night-adapted, eye has the potential to beat even that. It is a tragedy that so few can now exploit this wonderful capability to enjoy the greatest show on Earth, laid on free every night, as a God-given privilege.

Visibility is defined according to *relative*, not absolute, brightness. The ability to distinguish two features with brightness A and B depends upon their *difference*, not their degree. We can calculate a figure for visibility from the ratio of difference to sum: $(A - B) \div (A + B)$. If, say, both A and B simultaneously double, the visibility of the fainter feature is unchanged.

Suppose A is the brightness of an oncoming car with headlights, and B is that of an unlit background. Their difference is great. It's easy to spot an oncoming car at night on an unlit rural road. Now add a streetlight. It does nothing at all to brighten the car, but it does brighten the background, and consequently *diminishes* the visibility of the car.

Another mistake is to fit brighter headlights. When A is already much greater than B, making A still bigger has no significant effect. What you *will* do is increase glare, making everything else on the road *less* visible. New brighter halogen headlights actually diminish overall safety, with only a modest benefit for the owner.

Since they are otherwise totally unlit, there can be no doubt that street lighting makes a pedestrian in the road more visible to a driver, and thus reduces their chances of joining the annual casualty statistics. This would seem its only justification. There are two objections, however.

First, street lighting is not very effective in making pedestrians visible, because their background is equally illuminated. A better way to improve their visibility would be for them to wear high-reflectivity clothing, as is provided to all those who must work in the road these days. In any case, it is far easier to help a pedestrian avoid a car than the other way around, which is why a kerb-separated pavement and signed crossings are vital.

Second, we might call into question whether it can *ever* be truly safe to mix motor and pedestrian traffic. The lighting argument disappears where they are separated. As a step back from complete separation, many more properly signed crossings could be provided, removing the need for a walker to ever find themselves in the road.

It seems only reasonable that the full cost of pavement and crossings be met by the motorist, through road-use charges, and not the pedestrian. After all, it is the driver who poses the threat to the pedestrian, not *vice versa*.

In Britain, it has (with neither debate nor mandate) now become policy to extend lighting to areas *void* of all pedestrians — even rural motorway.

This is quite unnecessary, and gains only pollution.

Retroreflectors — or "cat's eyes" — alone can enable a driver to see the boundaries of the road ahead to the full extent necessary. Colour can discriminate layout, including exit lanes and junctions.

Both "cat's eyes" and high-reflectivity signing become *less* visible with overhead lighting. As we have already noted, so do other vehicles.

Street lights are neither cheap to install nor cheap to operate, despite the efficiency of the discharge lamp. Unlike the lights in your home, which operate for an hour or two each day, street lamps are all lit twelve hours per day, on average, all year round.

They may also intrude on your home. Our first love-nest was a newly built house with a garden barely big enough for the neighbour's cat to foul. It was nonetheless dear to us. We were profoundly dischuffed to return home from holiday one day to find a street light erected right in the middle of our little lawn. Apparently, regulations dictated a lamp every so many metres, but did not burden the local authority with any duty to even let us know.

Our neighbours found a lamp right outside their bedroom window.

Each street light costs around £3,000. None of those erected near our home could ever be considered necessary. Next door was a freight yard lit up all night like a football pitch for a cup final.

There are now around 7.5 *million* street lights in Britain, each consuming about 100W. That tots up to about 750mW — a significant proportion of the output of a typical power station — costing over £35m each year. Even if each lamp lasts twenty years, the annual replacement bill will run to over £1bn. Maintenance and other overheads will bring the total cost closer to £2bn *per annum*. Given that much of the stated justification is road safety, all this must be considered yet another uncounted cost of motoring.

Aside from its direct effect upon our environment and pockets, street lighting also has a significant impact on nature. Intrusive illumination can inflict damage on the local ecology, interfering with the natural behaviour of mammals, birds, insects, and even fish. Concern is growing over the effects of light pollution on wildlife.

Some creatures can adapt. Many cannot. This would be one thing if they all lived independent little lives, but they don't. Some ecological systems are being seriously disrupted. The effects will gradually spread, and on varying time-scales. As with our climate, a grand uncontrolled experiment is in progress, whose outcome may prove tragic.

When you break your car, you can get it fixed or buy a new one. When you break an ecology, it is probably gone for good.

Noise is a problem that arises with all transport corridors; air, rail, and road. All aviation noise emerges from aircraft engines, which have become much quieter over the last decade or two. Railways are mostly permanent fixtures that have been in place for over a century. Anyone moving in nearby knows the choice they are making. New railways, such as the 300kph Channel Tunnel Rail Link (CTRL), have been sunk in tunnels when passing through residential areas.

One difference with a road is that the noise is continuous and unrelenting. Most of it is caused by air squeezed between tyre and surface. Concrete surfaces are significantly worse than tarmac, partly because the surface must be corrugated to provide sufficient grip. This creates little shafts that act like organ pipes. The tyre squeezes air down them causing sound. Tarmac avoids this, while providing better adhesion and better drainage, especially when porous. Concrete has been widely used because it is cheaper, but is now deprecated in deference to safety, as well as noise abatement.

A skim of porous tarmac can be added to a concrete road, with immediate and considerable improvement, and at little cost. In The Netherlands, any road carrying more than thirty-five thousand vehicles a day *must* be skimmed, at least.

The degree of noise created by a road depends heavily on the speed of its traffic. Measures to counter excessive speed can often be justified on grounds of noise abatement alone, quite apart from safety. Speed cameras, bumps, and chicanes, have proliferated in recent years, in Britain. Though subject to occasional rebellion by a minority of drivers, who consider such restraints an infringement on their liberty, they have been as successful at reducing noise as they have at eliminating casualties.

Another way to abate road noise is to saturate it with traffic. While congestion is hardly desirable, it does at least relieve the effects of speed.

Road noise is a growing public nuisance, with complaints rising consistently. The American Environmental Protection Agency has proposed it as the "leading source of community noise", understating the obvious for many. The European Union estimates eighty million of its citizens suffer unacceptable levels.

Wildlife is again seriously affected, particularly birds in their breeding habits.

A common mistake is to infer that a cost to living near a road is exacted only upon urban residents. Rural areas are not spared exposure either to road noise or light pollution. Highways that once were quiet, and lit only along urban sections, are now noisy and fully lit. A concentration of services and amenities away from villages and small towns, together with "rat-running" to avoid congestion on trunk roads, is increasing traffic far from the madding crowd. The rural idyll — the aspiration of so many retirees — is now under serious threat from creeping urbanization.

With regard to health, road noise is of the worst possible kind, being both continuous and 'white' (possessing a broad frequency spectrum). Continuous unrelenting white noise has been used for military interrogation, which hardly recommends its acceptance within a community. It can break down the will to resist and render people submissive and subservient. It can also cause fatigue, reduce working efficiency, and interfere with communication and sleep, giving rise to stress. Physiological effects are also known, including the raising of heart rate and blood pressure.

Road noise is believed to translate into real disease and hospital admissions, though establishing a causal link is extremely difficult.

Beneficial it ain't.

Out of sight, out of mind

We have seen how roads can divide communities, remove the pleasure, efficiency, safety, and ubiquity in both walking and cycling, and incur light, noise, and air, pollution. It can threaten health and destroy architecture and landscape. A once vibrant urban centre, rife with culture, society, and commerce, can degenerate into ugly, even dangerous, destitution. Many towns are now without a beating heart.

Suburbs were borne in response, as a retreat from a hostile environment, but inevitably bring about a great rise in the transport demand. While the 'burb' is disassociated from the hubbub, the town is blighted even more.

There is an answer. Build a high-capacity network, and bury it.

While the humble bicycle remains the cheapest, most ubiquitous, and (when isolated from motor traffic) safest, means of transport over distances just too far to walk (say, from one to ten kilometres), underground rail conduits are ideal for longer journeys, and can offer the greatest comfort and speed.

Only rail can justify tunnelling underground, as it affords the highest capacity and the greatest safety. A very high initial cost can be set against a greater return sustained over a longer period. It calls on us to invest on the behalf of many generations to come, as our ancestors once did. They were bold and visionary. Within fifteen years of the first commercial overground passenger railway, planning of the first underground lines began.

The payback has been incalculable.

London without its Underground network is as unimaginable as the labour, organization, and commitment, it took to build. A quarter of a million people converged on the capital each day by road in the 1840s. Chaos reigned. That commuters, by the thousand, were later willing to endure smoke and soot in open carriages underground is testimony enough.

The world's first underground railway — the Metropolitan — opened in January 1863. As a fully commercial venture, requiring only permission from the State, it was an enormous success. In a city then of three million, nearly twelve million journeys were made on it in its first year of operation.

Great innovation, like great art, can always be traced to an individual. We owe the London Underground to a City of London solicitor called Charles Pearson. Such people suffer dearly in convincing lesser minds, and his victory was only partial. Pearson recognized the need for a grand station in the heart of the city with underground connections to the main overground lines, which terminate all on the periphery.

Because London still lacks such an amenity, connections through it remain a nightmare to this day. In contrast, Berlin will soon have a grand central station, where one can arrive on a "ligne à grande vitesse" (LGV) at the hub of an urban underground network. Such an amenity also provides a powerful architectural focus, around which form can follow function.

Sadly, the CTRL London terminus will again lie on the periphery. Travellers face a slow and arduous struggle across the metropolis.

Sadder still is that so few cities possess any underground rail system at all. Many are denied rail travel altogether.

So what?

Three factors underlie problems local to surface transport. Far the worst is the internal combustion engine (ICE). Toxic, as well as greenhouse, gas emission is inevitable, often accompanied by VOC-ridden particulates. While this effect can be ameliorated, through "lean burn", catalytic conversion, particle traps, and cleaner fuel, only electric traction can offer a complete solution. The hydrogen fuel cell, fuelled via electrolysis, has the potential to eliminate emission altogether, except for plain water.

The next factor is the rolling wheel — the dominant source of noise, which grows exponentially with speed. While wheels remain essential, their constitution and track differ with transport mode. Rubber tyres rolling on a concrete surface generate considerably more noise because of the "organ pipe" effect they create. While this is mitigated by the use of a porous surface, it grows worse with rising vehicle weight.

The iron wheels of a train are the equal of rubber tyres only at much greater speed. Trains also need fewer wheels per passenger or tonne-freight.

We shall neglect lighting as a factor since we have concluded that it is no more necessary to light a road than it is a railway track. Street lighting is only truly warranted where people walk and congregate, to promote a feeling of security. Using it to light the way for drivers is simply misguided.

While that misguidance persists, however, a high cost is incurred.

Last, and perhaps eventually the equal of the ICE in its damage to our health, is the promotion of indolence. While walking and cycling both constitute excellent exercise, reliance upon any form of personal powered vehicle denies the most common source of it — just getting about.

Even electric vehicles promote indolence, perhaps even more so than the car, as we now know it. Their design can be more flexible, leaving them even more convenient. Some have already become acceptable within a shopping mall, sometimes even within the shops themselves.

While such vehicles will, for some time to come, of necessity be smaller and lighter than the existing beast, there is nothing in principle that limits their size or weight. Hydrogen, fuel cell, and electric motor, may weigh a bit less than gasoline, ICE, and power train, but much of the mass of a SUV is given over to a chassis and body strengthened to carry more and better resist collision. Later electric products may develop the same way.

And they will run on rubber tyres.

The kind of transport we use and the nature of the habitat we construct for ourselves are two sides of the same coin. Villages and small towns had foci, connected by rail, and could be navigated on foot and bike. Sprawl can be served only by the car, which becomes essential for *every* trip.

Questions remain, over our future path, such as whether we might retain the liberty of owning and using a car, without being compelled to do so.

Can we have our cake, and eat it too?

5

Routes and urbs

5.1 It seemed like a good idea at the time

To a generation of men raised in the Great Depression, recently returned from a terrible war, and having endured years of austerity afterward, the idea of possessing their very own private car was irresistible. With peace, came eventually almost full employment, along with rising pay and aspirations. The relaxation of social mores regarding debt, and the arrival of new and attractive ways of borrowing, was all it took to start a tidal wave of car purchase through the 1950s and 1960s.

Once you have invested so much in a car, you have every incentive to use it, and to the fullest extent possible. The comparatively modest additional expense is no deterrent. Destinations, and most of all jobs, that were previously out of convenient reach, suddenly become easily accessible.

To men who had seen their war won by computer, radar, and nuclear weapons, and who were seeing other men venture into space, this was *progress* — unquestionably good, and 'unstoppable' anyway. Faith in progress was replacing the more traditional kind perhaps because of the hardship that remained, and the new threat of belligerent communism in the East. Technology offered to remove both, and soon.

Mass motoring was just part of the process.

Powerful lobby groups were able to exploit this popular outlook to convince the State to invest billions in new and improved roads. The railway, hog-tied by state control, was in no position to modernize and compete. The more people deserted it, and the more state investment went into the road, the greater the incentive to answer temptation, borrow the money, and buy a car. But most of all, it was a matter of personal esteem.

Arriving on a bike impressed neither your girl nor your boss.

Business interests were also to respond. The out-of-town shopping centre could offer a whole new experience and exploit lower costs to boot. From the quaint village pub to new national 'superstore' chains, business moved out of the High Street.

Big business was already aboard. Behind the car manufacturers themselves, were the oil companies, awaiting an unprecedented bonanza. Facilities and technology developed to serve a world war were ready and waiting.

Politicians had little alternative but to comply. Despite the enormous numbers already employed on the railway, car production was perceived as the great provider of new jobs, and the means of restoring past industrial glory. Consumption was as important as production; after the loss of empire, Britain and France, in particular, had to develop a strong home market. Taxes had to come from somewhere to repay war debt. Lastly, *something* had to be offered in trade for food and raw materials.

Business was still about manufacture, and freight transport was still vital. The railway was seen as obsolete— a burden rather than an asset. Even those who could still see its advantages remained bitter over its virtual monopoly. Road transport was regarded as the "free market" solution, despite its reliance upon public provision of the highway itself. With little or no analysis of cost or consequence, a new motorway network would be built, along with massive investment in major existing roads.

With no effective competition, fortunes were made in road haulage.

As people followed freight in their migration from rail to road, the cost of using the train rose. Services were cut and stations closed, lengthening the journey and increasing the incentive to buy a car, which grew more affordable each year. Holidays — the creation of the railway — were particularly appealing in the car, which brought new 'undiscovered' destinations, and the possibility of wider exploration on arrival.

Eventually, the railway, already undermined in its core freight business, suffered a collapse, with a third of it closing in Britain.

Anyone denied a car, by ailment, infirmity, or relative poverty, now faced an uncomfortable, slow, and expensive, journey on the bus, or a more dangerous ride on their bike. Women, especially those with children, were especially badly affected. Even today, running a *second* car is beyond the reach of many ordinary folk.

With railway closures came the loss of an effective primary transport network. Only the road network would remain complete. The irony of the idea that this was somehow progress lay in the fact that the railway not only used far less energy to move the same mass, and at far greater speed, it was also the only mode that could one day be fully automated. The vast majority of car passengers are also drivers, committed to their task, and unable to work, rest, or play. Costs, of every kind, rose.

The "modal inversion" between road and rail is now understood to have been a calamity that increased the cost of travel (measured in journey time) on *both* modes. On rail, journey times are lengthened by the increased distance to and from stations and greater delay in departure and connection. On the road, journeys are delayed by congestion. We now understand that congestion is limited only by the journey time available on a competing conduit that can carry more traffic without slowing. The only credible and proven system that can provide such 'broadband' competition is a railway.

The only way to reduce road congestion is thus to invest in rail.

With the rise of the car, a 'blind eye' was turned to the carnage it caused.

Many men returning from a long war would have been used to the sight of death and injury. It can be no surprise that they regarded the inevitable collisions as mere accidents. When dealing every day with machinery that was hardly designed with safety in mind, injuries were common. Safety standards were similarly low in pre-war industry. It was 'unmanly' to complain. You would simply hear that "*life* is dangerous".

Ensuing generations have become similarly inured because death on the road has been the backdrop of their entire life. A report in my local paper on a child knocked off their bike and killed merited no more than a square inch. No question was raised. It was simply accepted, as nothing particularly unusual — just another accident.

In Britain, it is said that the road has become safer, because the rate of death has fallen in recent years (from 8 down to 6.5/bn pass-kms, among all UK road users, 1994–2004). In response, we might first ask to what extent this is due to vulnerable pedestrians and cyclists being deterred from venturing out. We might then point out that such figures are still far beyond what is considered acceptable in any other context, in the rich world at least.

They equate with an airliner crashing each and every month.

Of course, a hundred times more passenger-kilometres (pass-kms) are travelled by car than by plane. But the question remains: if travel by plane were to rise to match that by car, would we tolerate the same casualty rate?

That we might come to accept such losses seems most unlikely.

Rail travel used to be similarly dangerous. As with aviation, casualties were concentrated into a small number of dramatic events — the currency of the mass media then, as now. Public outrage (and probably fear) caused a new science of "safety engineering" to be borne. In Britain, in 2004, not one rail passenger died as a result of an accident caused by train or track. It is not without significance that several died as a direct result of road traffic. One driver parked on a level crossing to commit suicide; another careered down a railway embankment, having fallen asleep at the wheel.

Death on the road is by no means limited to those inside a vehicle. Both walking and cycling are intrinsically safe, yet they have become by far the most dangerous ways to travel, solely because of motorized traffic on the road. For every billion passenger-kilometres travelled in Britain, in 2004, just five perished in a car. *Thirty-nine* died on a bike and *fifty-two* while walking. Such a threat surely will understandably deter many.

No one knows how many would have died *but* for deterrence.

Yet counting casualties utterly fails to describe this, most terrible, cost of the car. For each death, there is grief and loss to the family affected. Children grow up without a father or mother. Parents bury children. Some of these additional victims see their health damaged, through depression, for example. Families are destroyed. Futures are denied.

As I wrote, a little girl of four was reported mown down on the pavement while walking home from a nursery, holding her mother's hand.[1]

Her mother lost a leg.

Statistics reflect neither the meaning nor the after effects.

---

[1] The driver was sentenced to thirteen years in prison for driving an unsafe vehicle, and driving under the influence of drugs.

Perhaps the most serious generic consequence of this new and dreadful threat is that children have become indoor creatures, no longer safe to roam about and play outdoors. With walking and cycling so dangerous, they have also lost the freedom to exercise and travel independently. Their social, emotional, and physical, development is thus impaired.

So, we face epidemics in childhood depression and obesity.

After the carnage, comes congestion, as the second major cost of the car.

The Royal Commission on Environmental Protection, the Confederation of British Industry, and Friends of the Earth, make strange bedfellows, but they all agree on a *minimum* figure for the cost of congestion in Britain — around £20bn each year. This represents the cost in lost production alone, caused by delays caused solely by weight of road traffic.

It takes no account of lost business.

Road transport is inherently congestible because traffic is divided into small vehicles, with very low occupancy (around 1.2 passengers only). A safe interval must be maintained between them that grows disproportionately with speed. When access is not controlled — for example, by market pricing — then the relationship between speed and space acts in reverse.

As vehicle separation falls, so must speed.

The *marginal* cost (that imposed by allowing just one more vehicle access) of using a road must account for the effect on *all* existing users. Were a charge to be levied that reflected that cost, enough people might choose to use an alternative 'broadband' service to maintain an acceptable speed.

Without such an alternative, no charge can be effective.

As noted earlier, to reduce road congestion, we must invest in rail. The snag is that the investment needed to reverse modal inversion is enormous.

As a new generation matures, it is expecting to own and run, not just one car per household, but one car per *person*. Its members also expect to remain free to use their cars without restraint. They see congestion as a problem for the State, as almost the sole provider of road space. Few will ever see themselves as contributors to it. Few will have ever used what is left of the railway. Fewer still will see the railway as part of any solution.

We have driven up a cul-de-sac without knowing how to reverse.

All governments and major scientific agencies now agree that global warming is underway, and most, perhaps all, of it is due to our burning of fossil fuels. Agreement is similarly strong that the likely consequences are catastrophic. A dramatic reduction in global carbon emission — by perhaps 60% by 2050 — will be necessary to restore stability.[2] To be fair to newly industrializing nations, the already developed world must make an even greater reduction — of at least 80%.

Fossil fuels currently dominate energy provision, accounting for about four fifths of it globally, and even more in the US. Consumption breaks down into domestic heating, electricity generation, industry, and transport. In industrialized nations, transport overall consumes about a third of the total, and road transport a quarter.

The essential difference between the two main overland modes of transport is that road vehicles must carry their own energy supply, *i.e.* fuel.

Rail can use power generated, by any means, at fixed locations.

All other primary demand for energy can ultimately be met likewise. Sustainable sources, derived from either solar or geothermal energy, can replace fossil fuel in the generation of heat and electricity for homes, and power for industry. This will depend on a drive for greater energy efficiency — an economic imperative anyway. Carbon capture and storage is another option for large-scale power stations, if it can become proven.

Only the automobile must drink oil to survive.

The only alternative fuels advanced are hydrogen and a variety derived from crops. Neither are available in adequate supply. Biofuels will tend to drive up the price of food, since they compete for the same land. While fuel crops can be grown on otherwise unproductive land, they could only ever meet a small fraction of current demand.[3] Hydrogen can be derived in various ways, but the only sustainable means that might conceivably

2 The only international agreement regarding atmospheric carbon emission at time of writing is the Kyoto Protocol, agreed in 1998 and coming into force in 2005. It is to reduce emissions to 5% below 1990 levels by 2012. Given the level expected without the protocol, this represents a 30% cut, as of 2005.

3 According to one estimate, even in the US, which has much more arable land available *per capita* than Europe, only around an eighth of the gasoline currently consumed could be replaced with bio-ethanol.

match demand is electrolysis. This implies a massive increase in electricity generation, which, in turn, would have to be sustainable.

While the hydrogen fuel cell has the potential to power future automobiles, its affordability is a very long way off. Also, the entire fuel distribution system would have to be replaced, at gigantic cost. It will take decades to rebuild, after all the problems in the safe handling of hydrogen have been solved. Meanwhile, the proportion of fossil fuel carbon emitted by road transport (dominated by the car) has increased, from 19% in 1990 to 22% in 2005, in Britain. While industrial emission fell by 13%, and residential emission remained stable, that of road vehicles rose by 10%.

The car is the greatest obstacle to stopping global warming.[4]

There are other reasons why we might reduce a reliance upon fossil fuel. Olaf Stapledon, in his mighty future history *Last and First Men*, foresaw one such reason as long ago as 1930 : [5]

> "The economic life of the human race had for some time been based on coal, but latterly oil had been found a far more convenient source of power; and as the oil store of the planet was much smaller than its coal store, and the expenditure of oil had of course been wholly uncontrolled and wasteful, a shortage was already being felt. Thus the national ownership of the remaining oil fields had become a main factor in politics and a fertile source of wars."

He later describes a catastrophic war between America and China largely over the last great reserve, beneath Antarctica. As I write, Britain is beginning oil exploration off the Falkland Islands, in the face of Argentina.

Perhaps one war has already been fought over that last reserve.

For any nation to be dependent on a raw material which can only be found in sufficient quantity beneath foreign lands is unhealthy, to say the least. It invites conflict. The US is "addicted to oil", to quote President

4 Two other threats are from shipping and aviation, but only because of their projected growth rates and high emission per trip. Shipping uses particularly 'dirty' fuel and no filtration — a situation that can be addressed through regulation. Aviation is seeing most of its growth in relatively short journeys overland, which can often be made in less time using high-speed rail, as has been shown in France.

5 Chapter 1, Section 4.

George W. Bush,[6] and thus depends for its very survival on the Middle East, where the only adequate supply is to be found.

Europe, China, and India, are growing a similar appetite.

Insatiable oil consumption in the rich world has already precipitated war in the Middle East, along with the prospect of an imperialism that, while it purports to bring democracy, actually undermines it. The seeds have thus been sown for conflict on a far grander scale.

If our children are to know peace and security, some alternative to "guzzling gasoline" appears vital.

Our grandfathers walked or cycled to work and to school, and did manual work. We drive, or are driven, and toil mainly with our minds. America suffered the beginning of car dependency long before Europe, and so has a slight lead in the obesity statistics. While in 1960 a still worrying one in eight American adults was obese, now it is one in *three*. Britain is catching up fast with one in four already. One in six British children are now obese, the number having tripled while that of adults only doubled.

A change in diet cannot explain this since our bodies are normally able to burn off or eject any excess. In any case, people eat no more now than they did in 1960. The change *must* be due to physical indolence.

Children need to walk or ride to school and return to outdoor play.

The consequences of denying them the opportunity to safely walk, ride, and play, outdoors, will be epidemic diseases of the heart and circulatory system, diabetes, and a host of other ailments both major and minor. Mental health may be expected to suffer similarly.

Then there is the bill to pay in medical and personal care.

Longevity, which has risen in recent generations, may now decline.

We have not confined ourselves to our homes and cars *only* because of the physical threat posed by the busy road. For many of us, the great outdoors is not so great any more. The air we breathe is no longer fresh but tainted by oxides of sulphur and nitrogen. "Volatile organic compounds" embed themselves in soot that can find its way deep into lungs to cause cancer.

---

6 State of the Union address, January 31st 2006: "America is addicted to oil, which is often imported from unstable parts of the world."

While industry is responsible for some of this, the majority comes from the road.

Asthma in children has risen sixfold in the last twenty-five years. While pollution by road traffic cannot (yet) be proven as a cause, there is growing evidence that it is a factor. Children who do still play outdoors have been shown three times more likely to develop asthma where the air is polluted with ozone, which is already known to trigger and exacerbate attacks.

Internal combustion engine (ICE) exhaust also causes a cocktail of minor ailments, from fatigue and poor concentration to nausea and migraine.

Air pollution is not the only kind brought about by busy roads. Many more people each year suffer severe road noise as traffic levels grow. Continuous background noise can have a serious effect on us. While we can "tune it out" for a while, we rejoice when it stops. It affects our health, causing fatigue and stress, and is even capable of raising heart rate and blood pressure.

There is a reason we seek out and value tranquillity.

Lastly, there is light pollution. Most of us now live with that evil orange glow on the horizon, blotting out the precious stars at night. It is wrongly accepted that no road is safe to drive upon unless floodlit. The financial cost, in Britain, runs to perhaps £2bn each year. The real cost is greater.

All the glory of the night sky has gone the way of the wildwood.

None of these costs are due to the car *per se*, but to its overwhelming dominance over transport. Were the car simply an option chosen by a minority, there would *ipso facto* exist the liberty of choice. Modal inversion might not have occurred, allowing both rail and road to thus offer lower journey times. Road congestion could be avoided. Our dependence on foreign oil would be far lower, as would the proportion of carbon emitted by transport. With far fewer cars, driven far fewer kilometres, cycling and walking could return to the safety our ancestors knew.

There is no intrinsic reason why the liberty to own and run a car should end. The problem is entirely that the liberty *not* to own one has already ended, for most of us in the rich world. A substantial proportion of ordinary people cannot afford what has become a necessity. In Chapter 2, we showed how the true cost of running a car runs to over 40p/km (65p/mile). Few can afford to finance their car from savings or inheritance, especially while young, and so debt enters the household. Longer hours are worked and less is saved. All this weakens the family, often causing dissention and conflict. Working further afield, to earn more, takes parents away from children for yet more time, exacerbating the situation and worsening the traffic.

Once again, it is children who suffer most. It is fast becoming the norm for them to be deprived of both parents from dawn till dusk. While meeting the cost of the dwelling remains the biggest demand, meeting the cost of running two cars comes a close second. Some surveys have even concluded that a large and growing proportion of households spend *more* on their cars than they do on their home.

Cars don't hold their value like houses.

Whatever weakens the family, weakens the community. But the community is damaged directly as well. Motor traffic clears streets, not just of children playing, but also of neighbours getting to know each other and building trust. As amenities move out, towns become little more than dormitories, deprived of life and affection.

Finally, there is one more element of both family and community that is excluded from all benefit, yet suffers the cost, of the car — the elderly and infirm. As the baby boomers meet the final phase of life, many of them will be deprived of their car, on medical grounds. This will come as a gigantic shock to those who were first to come to rely upon it totally, *en masse*.

By then, there may be no shop, surgery, cinema, restaurant, or station, within walking distance.

Since the end of World War II, the dominant style of development has been that of car-predicated sprawl. In such an environment, there is little or nothing within walking distance. *Every* trip must be made by car, regardless of purpose. This is in stark contrast with the traditional town or village. These either evolved long before the arrival of mechanized transport, or were built to serve the factory, after the Industrial Revolution. Either way, everything was accessible on foot or bike.

In the industrial town, the bike played a greater rôle than the horse had ever done as it was far more affordable, and was thus adopted by the mass of ordinary folk. Things could thus become a bit more spread out. However, the real difference was made by the railway, which was built originally to serve industry, carrying freight between dock and factory.

At the start, the railway changed the lives of ordinary people less than one might expect. Passenger travel was too expensive to offer them more than an annual holiday. Work, school, and shops, had still to be nearby.

But the railway *did* begin suburbia. Those who could take advantage of professional work available in the city could now move out to "the country" and breathe fresher air. The pattern of land use changed, but not that much. Foci remained, since every new suburb grew around a station.

The railway merely added wealth and opportunity for some.

Throughout all history, the forms of transport available dictated the pattern of land use. Whatever the age, transport and the design of the human habitat are two sides of the same coin. With the arrival of the car, however, the pattern of land use has come to dictate the form of transport.

Again, we can see that the car itself is not a problem. Choosing to render it a necessity rather than a luxury, on the other hand, *does* create a problem. All the costs we have listed above make it one. Exclusion, loss of choice, social decay, pollution, ill health, congestion, longer journey times, excessive carbon emission, and wholesale carnage, all follow as a result.

Car-predicated sprawl is thus not a habitat suitable for humanity, anymore than a pit is for a bear. Nor is it in any sense necessary for modern life.

5.2 A city in a garden

The late twentieth century was not the first time our environment had suffered serious damage, at great cost to human life, health, and happiness. New tools had wrought havoc before, notably in the Industrial Revolution.

Just how idyllic life had been in village and field may be open to question; its quality in early industrial towns is not. Long days were spent slaving within oppressive factories. Sewage ran open in the street, causing regular outbreaks of cholera and typhoid. Aside from the physical deprivation, there was also profound moral and spiritual decay, brought on by a despair unmitigated by any tradition or culture, all of which had been lost.

Technological progress had annihilated a way of life that had stood for ten thousand years. No one will ever know how many failed to survive being thrown off the land. Many may have preferred death, especially had they foreseen the ugly nihilistic urban existence that awaited them.

Eventually, the causes of contagion would be understood, and physical health at least partially restored. But one other pestilence, born with the industrial town, remains with us to this day. Crime and antisocial behaviour arguably pose a threat to civilization at least as great as global warming.

There were some who sought to rectify the problem by the careful design of an entirely new kind of city — one that would reunite town and country. In 1898, Ebenezer Howard published his proposal for a "garden city", built within a circle, and with radial boulevards leading inward to a central park. It would be surrounded by a "green belt", peppered with allotments and small farms that would feed the population, whose size he suggested would number about thirty-two thousand. His plan for growth anticipated more such cities, interconnected by roads and railways.

The Garden City was a model adopted around the world throughout the first half of the twentieth century, but it had one critical weakness.

Howard had never envisaged the rise of the private car.

For those who sought liberty for the automobile, there was another, whose ideas accommodated it. A Swiss watchmaker's son named Charles-Édouard Jeanneret, better known as "Le Corbusier", espoused the notion that both home and city were no more than a *machine* in which to live.

Unfortunately, he became an architect, and his crimes were many.

His *Villa Savoye* was designed around the car and epitomizes the flat-faced, square-windowed, straight-lined, ugliness of 'modern' buildings that were to curse generations to come. Worst was his hideous 1922 vision for a *Cité Contemporaine* of three million inhabitants, who were to be incarcerated in sixty-storey skyscrapers to free up space on the ground below for the car. As the author of the "Plan Voisin", he proposed bulldozing vast tracts of Paris in order to realize this nightmare. To our eternal relief, it was met with scorn, and Paris survived intact. Sadly, his ideas were later adopted in Brazil, the Soviet Union, the USA, and Britain.

To his credit, Jeanneret could see the fate that was befalling the city, and asked the perpetrators to help address the problem :

> "I saw the heads of the Peugeot, Citroën and Voisin Companies and said to them: 'The motor has killed the great city. The motor must save the great city. Will you endow Paris with a Peugeot, Citroën or Voisin scheme of rebuilding ...' "

Peugeot thought it too risky, Monsieur Citroën "did not understand", but Voisin decided to back him and gave the plan its name.

High-rise apartment living can be fine for the middle class (including planners), who travel often and deposit children in a boarding school. It was rather less appropriate for the mass of ordinary people, especially those with a family. Corbusier, middle class and childless, might perhaps be forgiven for such an oversight. His many followers who rebuilt war-damaged and dilapidated urban heartlands in Britain and elsewhere in the 1950s are less easily absolved. In Glasgow, for instance, where tower blocks had replaced even more wretched tenements, four in five children lived above the fourth floor. After growing up with nowhere but a car park in which to play, delinquency is hardly surprising.

To summarize, Corbusians believe in concentrating people into *high-rise* blocks of standardized 'cells', strongly *zoning* land according to its use (residential, commercial, industrial, *etc.*), and laying only *roads* for transport. Neither the social nor the practical consequences — not even the area needed for parking — seem ever to be properly considered.

Thankfully, there are at least some in the planning community who share an awareness of them. For example, Peter Hall commented : [7]

> "Le Corbusier ... represents the counter-tradition of authoritarian planning, the evil consequences of which are ever with us."

The effect on urban planning in postwar Britain was to see tower blocks erected, while traditional streets were bulldozed to make way for multi-lane roads and (eventually, multi-storey) car parks. The results were social decay, to the point of unrest and riot, a huge waste of resources (most tower blocks have since been demolished), severe pollution, and chronic congestion.

We are all living with the aftermath. As the towers came down, they were replaced with suburban sprawl — once again, predicated on the car.

---

7 *Cities of Tomorrow*, 1988, p.5.

All the advantages of traditional town life were lost in both tower block and suburb. Community had evolved dependent on proximity within a population of moderate density. Town and village bound people to place, and endowed both with a strong identity. Block and 'burb are nowhere.

In a social sense, they do not exist, and are disowned.

London was saved the worst by a few who saw it coming. One who fought 'modernism', and the demolition of pre-car urban architecture, was Jacques Tati, the French movie maker. *Mon Oncle* at once celebrates old-town life and ridicules what was to replace it — hilariously, I might add.

Thousands head for France each year to enjoy what survived.

The process of industrial revolution is still running, as we continue to explore the extent to which machines can do the work. I remember well a neighbour who had only just become fully qualified as a pattern maker when his entire profession ceased to exist, almost overnight.

Computer-aided design and fabrication had rendered it obsolete.

In just the same way, we continue to struggle with the consequences of our changed habitat. City centres have suffered particularly badly, as people escape to the suburbs to avoid rising crime and antisocial behaviour, and the insecurity it engenders. Many centres face economic stagnation and even death as business and amenity move out. Decay fosters further decay until those left behind start to destroy what remains. Those safe out in the 'burbs can only watch ruefully, as the smoke rises above the ruin.

One Walter Elias Disney watched, but determined to do something.

Disney grew up partly in the country and partly in the city. He regarded his time in the small Missouri town of Marceline as the happiest of his life. Many of the vivid and enduring images he was to bequeath us were inspired by his experiences then. "Main Street USA" in Disneyland, and the Magic Kingdom at Disney World, was inspired by his recollection of Marceline. He believed in small-town life, with access to the countryside, and was never to be truly happy in Chicago or Kansas City.

The smoke he saw rising over Los Angeles one day in 1965, from his home in the hills, was from the infamous Watts Riots. True, the trigger was suspected racism, but a California commission was to conclude that the underlying cause was high unemployment and inferior living conditions — in other words, social and urban decay.

It was perhaps typical of Disney to act rather than pontificate. He is recorded saying much about what a new city could offer, but little as to what the problems with existing ones might be. As he had done with entertainment throughout his life, he would begin with a "clean sheet" and no presumptions. He would experiment.

The single criticism he is recorded making of the city, as it was, had regard to the car : [8]

> "I'm not against the automobile, but I just feel the automobile has moved into communities too much. I feel that you can design so the automobile is there, but you can still put people back as pedestrians again, you see. I'd love to work on a project like that."

And work on it he did; tirelessly, as usual.

According to his biographer, Bob Thomas, Walt "wanted to combat the tyranny of the car" having "witnessed how the automobile had changed a sunny, attractive town into a concrete-paved, smog-choked metropolis." [9]

He was aware of the Garden City movement in England, from time spent there after the war making movies, such as *Treasure Island*,[10] and could draw from significant architectural experience; above all, with Disneyland, which had opened in 1957.

That Disneyland represented a major architectural accomplishment can hardly be doubted. The legendary developer and civic activist James Rouse called it "the greatest piece of urban design in the United States today" and praised its "performance in relation to its purpose".[11]

Someone who could pioneer an entire new art form (full-length-movie animation), build a major Hollywood production studio from scratch, and invent the theme park, was not easily daunted. This was a true driven genius who did nothing by half measure. Disney determined not just to contribute

---

8 Press conference in Orlando, Florida, 15th November 1965.

9 *Walt Disney : An American Original*, 1976, Chapter 27, p.333.

10 His production company was unable to extract money from a crippled postwar Britain, so they simply spent it there instead.

11 Inaugural speech at the Harvard School of Design, 1963.

something to an existing city, nor even to be content building an entire new city, but to build one "in a constant state of becoming". He would build an *Experimental Prototype Community of Tomorrow* (EPCOT).

It was to be rather more than a theme park.

Unlike the 160-acre Disneyland, whose surroundings deteriorated to what he called "a second-rate Las Vegas", this would be controlled right out to the horizon. His team were to have 27,443 acres in which to plan not just EPCOT itself, but satellite facilities too, including an airport, an industrial complex, and a theme park. The last was the only item one might describe as ordinary. It would replicate Disneyland. On the other hand, the airport was to pioneer new ways of handling baggage, freight, and passengers, and was connected by high-speed rail. The industrial park was to accomplish no less than a 'showcase' of American industry, inviting spectators.

Even the theme park was to be five times bigger than Disneyland.

There have been those who have questioned Walt Disney's character and motivations, but then he left rather a lot behind as monuments to what he believed in and cared about. He was a family man — a devoted father to his two daughters — who saw art as enterprise. The motivation behind Disneyland, as well as his movies, was to create entertainment "the whole family could enjoy," together.

Disney was never motivated by money, except as a means to enable his next venture. His long-suffering brother Roy would ever be expected to come up with finance. There were plenty of opportunities to sell his share of the business and comfortably retire. That was something he could not do.

What drove and consumed him was his need to achieve. Perhaps this was to serve an ego that was certainly inflated. Perhaps it was a need to reassert self-esteem that had been crushed in his childhood.

Whatever it was, it was not greed.

While his treatment of his staff could be brutal — he had little hesitation in sacking those he regarded as "dead wood" — and play he would the ultimate autocrat, he was no dictator. This was no Corbusier, who would seek to force people into a standard mould to achieve some ultimate notion of order. When he first made real money, with *Snow White*, he spent much of it establishing a new studio with an unprecedented quality of working environment. Yet more of it went paying generous bonuses.

His adventure in planning EPCOT seems to have been motivated simply by all the things in which he passionately believed, and to demonstrate their value for all the world to see. When he learned of his impending demise, he complained that he would be remembered only for a mouse. There is something noble in seeking to be remembered for serving your convictions.

Ultimately, what more can any of us achieve.

Without any evident religious commitment, Disney believed simply in human 'happiness', and wanted EPCOT to be "dedicated to the happiness of those who live and visit there." He was passionate in his belief in free enterprise, hence his desire to provide a "showcase for American industry". He also believed deeply in the power of *imagination*, especially when combined with technology and industry. EPCOT was to "stimulate American industry to develop new solutions to meet the needs of people." Like so many in his generation, he bought the idea that technological progress had and would continue to bring "a bright new beautiful tomorrow".

There was one other thing in which he believed — the railway.

The Disney family had arrived in Marceline by rail, and he would have grown up regularly hearing the three-tone whistle. He arrived in Hollywood by train. When the studio became burdensome, he retreated into a world of model railways, building a half-mile track around his house and naming the engine "Lilly Belle" after his wife. When she protested about the molestation of her flower garden, he built a ninety-foot tunnel beneath it. His hobby led to the Santa Fe & Disneyland Railroad.[12]

It would be easy to decide that he was just another "railway nut" who could not accept the loss of steam trains. But then he was equally enamoured of his cars, in which he took great pride and pleasure. He never opposed their use. To deny anyone the option to own and use one would have defied his fundamental belief in liberty. The truth is that he, along with his engineers could see the limitations and liabilities of excessive *reliance* upon the automobile and road transport. As a result, EPCOT would be "transit oriented", *i.e.* it would be designed around various forms of railway.

All that was needed was to bring the railway up to date.

12 Later renamed the Disneyland Railroad.

To quote Steve Mannheim, with regard to Disney and transport : [13]

> "... Disney was ... in a unique position to ... ask people to think about what the automobile had done to their communities. Frequently, freeways divided neighbourhoods physically and socially. ... In the EPCOT concept, the automobile was banned from the dense commercial center, housing and employment were in close proximity, and electric mass transit systems were always within walking distance."

EPCOT would retain one Corbusian feature: there would be zoning. Four "primary spheres" are allocated separate space within a radial plan. Business and commerce occupy the centre, followed by concentric rings accommodating first high-density apartments, then recreation (within a wide greenbelt), and finally low-density suburban family houses.

The entire plan is strongly *focal.*

Unlike Howard, Disney placed a sophisticated transport hub at the centre. This made use of the third dimension every bit as much as did the "cosmopolitan hotel and convention center towering thirty or more stories" above it. Directly beneath, inter-urban monorail met intra-urban transit. Both were to be purely electric, and so would be welcome right inside a hotel lobby.[14] One level below lies the "automobile thruway" and car park. On the final subterranean stratum lies a separate highway for trucks and their "loading docks".

Instead of goods being carried through busy pedestrian thoroughfares, as they have done since cities began, they arrive underground and move up vertical shafts. Cars are also kept away from pedestrians, affording a far safer and more inviting environment. The narrator in the Disney EPCOT promotional video could boast :

> "Here the pedestrian will be king, free to walk and browse without fear of motorized vehicles. Only electric-powered vehicles will travel above the streets of EPCOT's central city."

---

13 *Walt Disney and the Quest for Community*, 2002, Chapter 2, p.40–41.

14 They actually do this at Walt Disney World. You arrive within metres of the check-in desk. In at least one Disney World hotel, you can even look down into a restaurant from inside a monorail carriage as you pass overhead, apparently disturbing no one.

The entire area was to be enclosed — a precursor of modern shopping malls, except this one stretched over *fifty acres*.

It housed more than shops.

A highway was to run underground through the hub, with a one-way ring road running around the city centre and its high-density apartments.[15] Spurs led to each suburban 'neighbourhood'.

Everything had been carefully planned to account for the expected traffic load; the city was far less prone to the jams and delays inevitable with the traditional grid system.

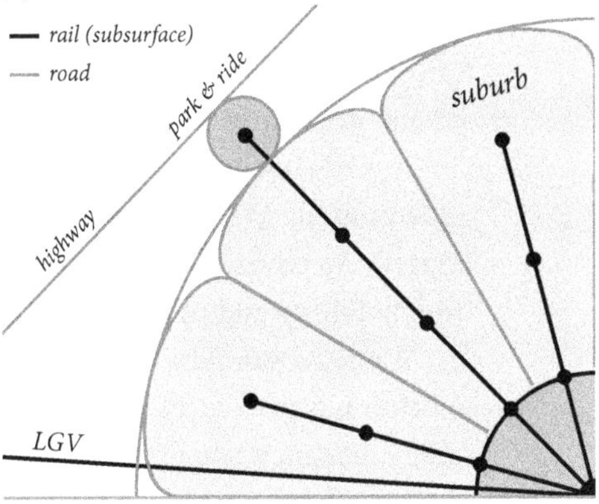

FIGURE 5.1 A 'transit-oriented' design for a city.

That load would be greatly reduced by the *choice* available. Radiating outward from the hub came the "WEDway people-mover".[16] The wedway was essentially an ultra-light rail system, driven by motors set into the track. This allowed a motor to be replaced without halting either system or train. Movement was continuous, though fairly slow at around 18kph (cycling speed), but offered a capacity of nearly 5,000pph. Boarding was via an ingenious rotating platform which synchronized with the vehicle. Otherwise, it was similar to the entertainment rides at Disneyland, where a five-eighth scale prototype ran in 1967. The people-mover can still be ridden today at Tomorrowland in the Magic Kingdom of Disney World.

Disney attributed a *prioritization* to transport modes. The pedestrian was king, and accorded ground level, along with bikes and "electric carts" (long before any such thing appeared on our streets). Electric "transit systems" were next, and allocated space overhead, without any of the deleterious effects apparent in elevated railways elsewhere, such as the one in Chicago Walt rode in his youth.

Cars and trucks went underground, though unimpeded in any way.

15 This road would act as a barrier to walking and cycling into town. A ring road running *outside* the 'neighbourhoods' (as shown in Figure 5.1) removes that barrier, without any obvious disadvantage to motorists. It may then also serve a "park and ride" depot.

16 WED Enterprises was a separate company used to develop the plans and tools for Disneyland and Disney World, including EPCOT. The name is derived from Walt's initials.

Neither the monorail nor the wedway were to be adopted elsewhere during Disney's lifetime. In fact the only take-up since, outside company property, has been in Las Vegas, where a system was recently extended. While "mass transit" and the use of clean electric propulsion may be right, these particular systems are clearly not.

It is obviously far cheaper to elevate a railway than it is to bury it underground. What is undeniably striking about a monorail is its modest impact. An elevated conventional railway is decidedly wider, robbing light from buildings and pavements.

While its slimness blocks less light, and arguably allows a monorail to better navigate a city, there is an adverse consequence. To achieve the same occupancy, you need much longer trains, and therefore much longer platforms, both of which add to construction cost. Long straight platforms are also not always easy to accommodate, however thin.

Any elevated monorail suffers from another problem. In an emergency, passengers cannot easily leave the vehicle and run to safety. They are forced either to wait for the arrival of rescue equipment or to use special slides, similar to those provided on an airliner. The only alternative is to build a walkway beside the rail, which would widen the structure, removing any advantage it had in narrowness, and adding yet more to construction cost.

The EPCOT monorail, pioneered by Alweg of Germany, is very quiet, which is attributed to its use of rubber tyres on a smooth beam, rather than metal wheel on metal rail. However, it is by no means clear that the latter are any noisier. Automobiles also roll tyres along a smooth surface, yet highways generate plenty of noise. Conventional elevated railways may simply be noisier because their track is typically neither smooth nor continuous. Such systems leave much room for improvement.

The only clear advantage of tyre-on-concrete is that it enables a vehicle to climb a steeper gradient, overcoming a major limitation of any conventional railway. But rolling a tyre on a concrete beam uses two to three times more energy than iron wheel on iron rail, as we have noted before.

Speed is also an issue. We now know that a conventional railway can deliver speeds comparable with aviation. While 130kph (80mph) has been claimed for an Alweg/WED train, safety and comfort must be called into question. The EPCOT system does not deliver a particularly smooth ride even at the much lower speeds at which it currently operates.

Not least of the difficulties encountered by such systems is that of switching track. Any apparatus for moving concrete beams around is both big and expensive. But the real killer is usually the fact that monorails are totally incompatible with the existing railway network. To establish a new standard medium, and compel writing off the old one, demands justification. Some critical advantage must apply, in, say, operating cost or speed.

No monorail has yet demonstrated any such property.

The wedway people-mover suffers disadvantages beyond its low speed and rather cumbersome name. Many small carriages will cost more than fewer larger ones, with no real advantage, other than that they suggest a little privacy. Embedding many motors in the track will cost much more than a few in the train. While this allows replacement without blocking the system, reliability can be achieved in other ways. Electric motors are, in any case, very much more reliable than a typical ICE.

For any transit system to be competitive with the car, it must visit as many locations as possible, implying that the cost of the track is critical. (The cost of laying roads is normally absorbed *de facto* in estate construction.) Given that the purpose of such a system is to serve the suburb, where there is more space, there is less advantage in elevating it. When tunnelling cannot be justified, visual impact and noise may be reduced by digging cuttings.

Tunnelling *can* be justified near the centre, where population density and land value rise steeply. Light and heavy services can meet underground.

What was undoubtedly right with the wedway was its automation, reliability, and continuous running. Even more so was the way in which it was thoroughly *embedded* in the architecture, both of each locality and of the city as a whole. The need for infrastructure was fully answered.

Since the system is isolated within the city, there is no issue of compatibility. Disney was wholly justified in considering radically new media.

He was a man impatient for what the future would bring.

Aside from its 'transit-oriented' approach, EPCOT also differed from La Cité Contemporaine profoundly in both scale and density. Whereas Le Corbusier dreamed of three million in his great glass towers, Disney intended a mere twenty thousand inhabitants.[17] Like Howard before him, he envisaged growth via the creation of new and similar cities, linked together by high-speed railway (monorail).

Habitation would always surround a hub containing both amenities and transport. Life revolved around foci, according to activity. You could choose between living in densely populated apartment blocks, around the centre, and the sparsely populated 'neighbourhoods' on the outlying ring. These were designed primarily for family life, with houses arranged around the edge of grassy 'commons', which served recreation, with playgrounds, tennis courts, *etc.* They also offered paths for walking and cycling to wedway station or school. No road interrupted them. No car or truck destroyed their peace and security. In the words of the EPCOT video narrator :

> "Children going to and from schools and play areas will use these paths, always completely safe and separated from the automobile."

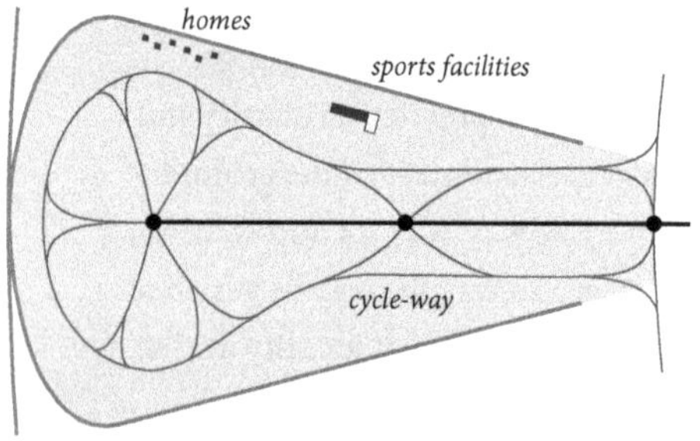

Figure 5.2 Logical structure of a neighbourhood.

Despite this, the design retained the freedom of residents to own and use a car. A road ran around each neighbourhood (Figure 5.2). Since each house faced onto it, both car and delivery van had access. Combining the ability both to use and to avoid the automobile seems so astonishingly simple, it is hard to understand why it is so seldom achieved.

Part of the trick seems to be to eliminate 'piecemeal' development.

The likes of Howard and Disney could plan for every need at the same time. Neither set out simply to make money. If they had, they would probably have repeatedly sought a modest parcel of land and built as many houses as possible on it. Transport and amenities would be someone else's

[17] Up to *sixty* thousand were spoken of in early discussions. As many as six million were also expected to visit each year.

problem. Those whose problem it was would have little choice but to allow housing to be built this way. Private investment is much more widely available and dependable than the public kind.

Without visionary and powerful leadership, integrated holistic development is unlikely ever to arise. Few find themselves in the position to try, as Disney did. When they do, the results do not always turn out as they intended. Sometimes, the individuals concerned are too radical or impractical, and require restraint.

Arguably, what is needed is just better understanding so as to elicit principles. In the next and final section, we shall attempt to draw some such principles from all that has been observed in this book. But Disney held out the prospect of something more. Each letter in his acronym was chosen deliberately. 'C' was for 'community', not 'city'. The 'E' in EPCOT meant something too. He did not pretend to have all the answers we need, so he proposed that at least one city — his — be *experimental.*

Perhaps this idea was his greatest contribution, at the end of the day.

Sadly, the man a nation had grown up revering as "Uncle Walt" died of lung cancer, just after his sixty-fifth birthday, in December 1966. He had long been a heavy smoker. Without his leadership, vision, conviction, and sheer determination, the EPCOT that was his final dream would not be built. Only a new theme park, under the same name, would arise, intended to reflect some of the key ideas in the city plan.

In the theme park, like the city, people could walk, or ride light electric carriages, free from the "fear of motorized vehicles". The many shops and exhibits are serviced via underground roads, loading bays, and lifts. Trash is removed by subterranean channels that lead to an incinerator. Only the almost silent electric monorail passes through, but overhead, causing no disturbance, threat, or obstacle. Its colourful movement actually adds life to the scene, which is one of great beauty, since the space that elsewhere would be dominated by roaring traffic could be fully landscaped.

The overwhelming impression is that of an idyllic village from a time where technology and reason have finally brought peace and prosperity.[18]

---

18 No, they did not pay me to write this.

Though the city itself was never built, its plan succeeds in demonstrating that we can aspire to something better than we have. It represented a bold attempt to at least *ask* the relevant questions and explore some answers.

Certain issues emerge as central.

Throughout history, urban planning has required walking the fine line between serving the individual and serving the community. Grant the individual too much and many, perhaps even the majority, may become excluded. Pay too much heed to the communal and the individual may starve. We are all individuals, but we also share a need to belong.

Like the original garden city, EPCOT made another point — that if you want to achieve something better, you must accept *design*. This implies granting authority to someone who will influence how, and where, you live. For most of us, this smells of dictatorship, though the fact is that we already surrender such authority. Wherever there is civilization, there will always be charge established over development and land use. It was precisely this authority that saved Paris in the 1920s from Corbusian destruction.

Like Howard before him, Disney strove to achieve liberty, in which he deeply believed, through *choice*, in style of home, work, recreation, and transport. But he began by recognizing that nothing could be achieved without complete authority over the design and planning process, and beyond, into the running and continued development of his community.

The problem of balancing the need for democracy and the need for such authority plagued him throughout. He was neither first nor last in this respect. It has always been much easier to get it right at the start than to keep it right over the decades and centuries to come. This is why the first element from the acronym was so important. We must first learn the rules which best govern building and running human habitat.

EPCOT related what some of the rewards for getting it right might be. The pedestrian might be king, and walk without fear or barrier. The commute to work might be free of frustration, even a pleasure. Every entertainment and amenity might be within easy reach. Children might walk and cycle in safety. With all transport electric, we could be free of pollution, at least where we live, work, relax, and play. Power generation could then exploit sustainable energy supplies, free of significant carbon emission.

We might become as free *not* to use a car, as we are now to do so.

5.3 The future is not what it used to be

How we got into this mess is less important than how we get out of it. The last thing any of us need as we face the morning traffic jam is a bunch of academics writing theses on how it happened. We can read about the underlying political failure later, once we've straightened things out.

Walt Disney understood this instinctively.

He appreciated that a proper solution would have to be *engineered*. A *design* must employ technology and satisfy constraints. These would require theory and calculation. Yet design remained aesthetic as well as quantitative. Disney saw the power in reuniting art and technology.

EPCOT debunked many a myth, such as the popular claim that the car is a "necessity for modern life". Another was the notion that the State alone can offer solutions to our problems. While his effort was not motivated by profit, Disney was not in the habit of losing money. His entire life had been spent making enough money fulfilling one dream so he could go after the next. This also had the effect of rendering things financially sustainable, and is why the movie studio and theme parks remain active to this day.

His last great project was no exception. In fact it took business into a new realm, where many distinct operations combined to sustain each other. In this, it remains unmatched in its vision.

Another myth to which he gave the lie is that the railway is somehow obsolete. First Japan, and later France, served to finally bury that one. The fastest way around Europe is no longer the plane, but the train. When carbon emission is properly accounted for, along with other 'external' costs to our safety and environment, rail will prove far cheaper also. Trains can float magnetically, for less noise and even more speed. Lastly, they can be fully automated, for the age when we have better things to do than drive.

Disney sensed that rail held the future when everyone else condemned it to history.

Finally, Uncle Walt challenged the idea that cities have to be *big*, with Corbusian millions somehow centrally concentrated. People have resisted that strongly, using first railways and later highways to create suburbia. They want space and to be close to open country, even to wilderness.

Only a tyrant would argue.

EPCOT was to be one of a number of separated settlements that are each closer to a small town than any typical contemporary city. Their size reflected the Platonic ideal. They would facilitate not just the rapid transit system he envisaged, but also *differentiation*, providing a choice of lifestyle.

Before I am arrested and charged with excessive Disney-mania, I should remark on a myth that his city *endorsed*. It had a centre, around which everything revolved. Everyone would work, shop, and play, there. There was a "grand central" station. It was 'hub-oriented'.

Transport networks, like computer networks, are often built around hubs, but these tend to cause more problems than they resolve. Ebenezer Howard placed his principal station on the edge of town. A great garden was to occupy the centre of *his* garden city.

Hubs are neither necessary nor particularly efficient. Instead, a *distributed* approach is possible. While it is beyond the scope of this book to dwell on technique, it can be done.[19] A homogeneous network sustaining distributed computation is termed a "computing surface" and has many advantages. For example, both machine and application can scale smoothly. Such a system can also be rendered "fault tolerant" much more easily.

What computer architects can do for information appliances, someone can surely learn to do for our habitat and transport system. The second invites the first to be focal in nature, but the net should spread wide.

The most intriguing cities have evolved to offer many, diverse, foci.

Perhaps we can evolve what we might never design.

For over a century now, the regulation of development has risen, along with prescriptive planning for new towns and cities. One early aim, in Europe at least, was to prevent cities growing out of control until they destroy wilderness and countryside, and "carpet with concrete" an entire region.

Any fear that a distributed approach to building our habitat would have such an effect could easily be ameliorated by early designation of areas

---

19 Homogeneous distributed computation was a research activity of mine for many years, with regard both to computing in general and to algorithms inspired by biological evolution.

we value, whether for their wildlife or beauty, or their alternate uses, for recreation or agriculture, or perhaps simply as wilderness.

Let us call it the 'net' — a distributive/focal habitat.[20]

As Disney suggested, we might set a limit on the scale of any development, forcing a new town to be built and added as the population grows. When the population falls, we can simply demolish those places we least like. Locations on the net could be large and small, urban and rural, in character.

Incidentally, another myth is that rural areas, with a low density of population, are spared the cost of the car. Nothing could be further from the truth. Rural lanes, with their varying width, and visibility impaired by bend and rise, have become prone to some of the worst carnage. People denied access to the car are often effectively disconnected from amenities altogether, with very little choice of work. If anything, the effects of the rise of the car are *magnified* where population is sparse.

The British countryside all but died with the loss of rural lines connecting villages to the national rail network. Viability of these lines was judged in isolation, but any transport network must be valued holistically. The loss of one destination denies a journey there from *every other* location.

Closing a third of any network significantly devalues the remainder, bringing about further closure. Recursion can end in collapse.

Wherever people live, they are surely entitled to basic infrastructure, including transport. This is presumably why no one has ever proposed closing a rural road. Obviously, there must be some limitation. We cannot all be expected to pay for someone who chooses to live alone, miles up a mountain. Charges can still relate to costs, but there must be real choice in lifestyle, including that of a reasonably small and remote settlement.

If we are to question the viability of the smallest settlements then surely we must also question that of the largest ones. Despite a century of attempts at "urban renewal", there are few, if any, large cities free from social decay. There will always be the feckless, the drunken, and the violent, but we must ask why are they the problem in a large city they never were in a village or small town. Parts of many cities are now denied us because of antisocial behaviour. They belong to no one, and no one belongs to them.

It can begin when an area is abandoned outside office hours.

---

20 A physical counterpart of its virtual namesake, sharing many of its properties.

Perhaps it is time to decide that the city itself, as we have come to know it, has failed. Perhaps it has simply failed to evolve along with us.

Cities originally arose as centres of government, religion, and finance, employing mainly a professional élite, as many still do today. Later, a new breed arose around heavy industry, and employed *en masse* mainly unskilled or semi-skilled labour. Very different patterns of development emerge in each case, but neither offered much of a life for the majority of ordinary people, forced, as they were, into servility.

Arguably, we neither want nor need such places any more.

Cities might well revert to something recognizable to our ancestors before the car. They may be much smaller and abandon 'zoning' to favour the mixing of residence, workplace, and amenity, which would reduce the pressure on any integral transport network, whatever its mode.

Though Disney's city was never built, his motivation and aspiration lived on and were shared by others. Just to the south of Walt Disney World, the company he founded built a town called Celebration, on so-called "New Urban" principles. These in many ways reflect what EPCOT sought to accomplish. They can be summarized as follows :

— each neighbourhood has a clear centre, complete with "transit stop"
— most dwellings lie within a 5min walk of the centre
— sufficient workplaces and amenities are arranged around the edge of each neighbourhood
— schools and playgrounds are safely accessible on foot
— parking is restricted to the back of each house, accessible via an alley
— streets are narrow to slow traffic, and connect so as to disperse it
— walking and cycling are supported and promoted.

Judging by property price, these ideas, and the environment they help to create, are proving popular. Once again, we are forced to wonder why such simple changes cannot be widely introduced.

The principal costs of the car are frequently addressed by the media in Britain. Even the rising obesity epidemic has now been publicly linked to the dominance of the car. Yet they generate only a deafening silence, in response. There is thus far no strategy for comprehensive change.

We must content ourselves with tinkering.

A 20mph (32kph) speed limit in residential areas, and near schools, is at last being spoken of, along with the introduction of road-space charging. Here we come to the last two great myths hopefully exposed in this book.

We saw in Chapter 2 that charging for road access cannot be effective unless a 'broadband' alternative is available. On both sides of the Atlantic, the railway is rarely now an option. Too many lines and stations have closed. Too little investment has been made in development and modernization. Without such an alternative, the only aim any charge could credibly have is to deter people from travelling at all, which is hardly helpful.

Almost every workplace, school, and shop, is now beyond a walk.

Even in the presence of an alternative, the charge will have to be determined by full market pricing, with regard to every aspect of demand. As was explained in Chapter 2, the very same charge that delivers the optimal return to the investor also offers the best value to the customer. There is no need for any transport system that depends upon heavy subsidy in order to persuade its use, which brings us to the final myth.

The bus, particularly the diesel variety, is merely what you provide as a last resort when confronted with the political unacceptability of social exclusion. In other words, if you have had no plan at all for transport, aside from road building, a bus is all you can offer the many who are denied a car.

Any idea that the bus represents a cheap and environmentally 'clean' alternative to the car is no more than a myth. The bus is both the slowest and the most expensive form of motorized transport yet invented. It even remains hideously expensive *after* unprecedented levels of subsidy. The air pollution it causes precipitates cancer, asthma, and many a minor ailment.

The bus is best seen as testimony to failure in transport provision.

Not only is the bicycle faster, *walking* is often faster, once the (often unpredictable) wait at the bus-stop is taken into account. Buses introduce a serious hazard to both walking and cycling, helping to deter both (to their operator's advantage). They cause severe wear to roads, for which they do not pay. Worst of all is their inability to serve sprawl — the prime cause of the inequity the bus is called upon to resolve.

Until recently, a valid argument in favour of the bus was that the elderly and infirm may not be fit enough to walk far or ride a bike. There is some circularity here. Many would not *remain* so unfit were they to walk and cycle more. Nevertheless, a minority would still persist in need of something else. A revolution in battery technology has now produced a variety of solutions, most of which could share the bike lane. Light electric vehicles (LEVs) include the electric bike, carriage, and Segway[21].

Some of these are in dire need of regulation — 12mph in a crowded supermarket is not on. But their appeal is growing.

We are left with no justification whatsoever for public provision (direct or via subsidy) of bus services, or even for allowing them into town.

While the purpose of this book is merely to document the cost of the dominance of the car over transport, and not to advocate any particular alternative, it does not seem reasonable to part without making any remark concerning how we might begin reversing out of our cul-de-sac.

Here are some principles that emerged from our consideration :

- both aesthetic and quantitative *design* are essential
- *foci* render transport efficient and promote community
- *distribution* denies congestion and promotes diversity :
  - no hub or 'centre' is truly necessary
  - growth can be achieved by extending the network, not the town
- mixing workplace, amenity, and residence, reduces travel :
  - zoning increases it, and no longer serves any purpose
- walking and cycling promote health and can satisfy most needs :
  - LEV can substitute for bike, and might use the same causeway
  - reliance upon foot, bike, and LEV, dramatically reduces 'land-take'
  - safety demands the separation of footpath and cycleway from road
- electrified rail offers the highest capacity, the lowest cost,[22] and the least environmental impact, for both inter- and intra-urban transport
  - between cities, rail also offers by far the greatest speed
  - within the city, rail requires the least street width (aperture)
- peripheral "park and ride" depots can provide for modal interchange, between car and bike, road and rail, and between light and heavy rail.

All these apply to how we construct new habitat, not just systems of transport. They go a little further than New Urban design.

---

21 'Segway' is a registered trade-mark of Segway Inc. of New Hampshire, USA.

22 Assumes all external costs, and returns over the full term of the investment, are counted. In comparison, the track cost of the bus is usually *dis*counted.

Left open is the question as to whether cars are permitted into residential areas or not. Here, there seems room for variation. Residential development that excludes the car can offer homes at a higher density and thus lower price. We have seen how urban design can also admit the car, without compelling any dependence upon its use. Regardless, it seems sensible and reasonable that cars belonging to visitors remain at a park and ride.

It is certainly possible to plan and build new habitat that is infinitely more appealing than car-predicated sprawl or a town blighted by traffic. Neighbourhoods can grow into vibrant communities, free of crime and antisocial behaviour. A sense of ownership, association, and responsibility, for a place can grow, fostering robust security. Children can enjoy far greater liberty and can develop properly through mobility and independence.

With the barriers imposed by roads removed, and the land they consume returned, piazza and park can arise in a fully landscaped environment.

The problem remains of how to get there from where we are. Although it might seem hopeless, given so much investment in road transport and sprawl, there is a lot that can be done, without wholesale demolition.

Arguably the simplest, most affordable, and most immediately beneficial, policy is to start building safely separated cycle/LEV lanes. This implies taking road space away from the car, including that currently used for street parking. Restraint, and possibly outright exclusion, of motor traffic from town centres, and in proximity to schools, can then follow.

Where they have been built, park and ride depots have proved very successful, for both passengers and freight. There seems no reason not to extend the idea to vehicles used for services. Even the police, in some towns, have switched many of their patrols successfully to cycle and Segway. Tools, and specialized equipment, can often be carried on a larger breed of LEV.

To cover longer distances, especially where shopping and goods are carried, light, and ultralight, electric rail systems appear desirable. These offer a clean, quiet, way in which to link park and ride depots to the town centre, and to each other. Such depots may perhaps form foci for new satellite development, pre-integrated with the transport network.

As alternatives become available, all subsidy, including the meeting of of track costs, can be denied the bus. Very few bus routes will then remain viable. Road space thus released can be used for cycle, LEV, and tram. The air can be clean and safe to breathe once more; cyclist and pedestrian can be free of fear and threat. Café culture can thrive in streets that are calm and quiet, and where there is space for plants, trees, and perhaps art. The town centre might become a pleasant place once more.

Population, business, and employment, may grow, as a result.

It will not be as easy, or cheap, to diminish reliance upon the car amidst sprawl. By definition, such places lack any workplace or amenity that can be reached by any means other than a car. About the only possibility is to *establish* a focus. This will be easy enough if a vacant lot of sufficient size becomes available. Commercial redevelopment is lucrative, and it might allow much higher prices to be paid for existing property than the market would otherwise allow. There seems no need to trouble the taxpayer.

Last but by no means least, the railway can be rebuilt and modernized to effect a complete broadband network, reaching every destination once again. Road congestion may yet allow facilitate re-inversion. As the cost of road travel (dominated by journey time) rises with each year, the task of making rail competitive becomes easier.

Recall that this would benefit both rail *and* road user.

Once all this is in place, leaving most people free *not* to run a car, then, and only then, might regulation require motoring to meet its full economic and external cost. This must include the cost of making the car safe, both for its passengers and for those making use of other modes of transport.

The liberty to maim and kill pedestrians, cyclists, and even the occupants of a train, must surely end.

It is easy to see why some see the State as the sole provider of transport. After all, cars rely on roads that are predominantly provided by the State, all around the world. In most countries, railway track is also funded from state coffers, generously filled by taxpayers, willing, unwilling, or just unknowing.[23]

The State has arguably been a little too keen to provide rather too many services. In Britain, in the light of rising public aspiration, taxation has risen as high as public tolerance will permit, and perhaps beyond. As politicians

23 On turning from employee into contractor, a fellow programmer noted in his blog (with appropriate expletives) his profound shock on writing out his first cheque to the taxman. One wonders just how long the present level of taxation would last if the majority of the population paid their income tax themselves, rather than leaving it to their employer.

try to cut costs, services whither. Without any competition, government agencies also tend to become, unsurprisingly, uncompetitive.

Many national rail networks are being returned to the private sector, to their increasingly evident benefit. We can expect road provision to go the same way. Road and rail will then compete as they did once before, only this time with a difference. Rail can hope to be free of excessive state regulation, obsolete technology, and an uncompetitive monopoly culture. In short, it can become as attractive an investment as it was in the beginning.

There is no fundamental reason why transport should not be provided by the market, at a market price.

Congestion is a direct consequence of doing otherwise.

We are living in a topsy-turvy world. The State is doing, at great expense, what it does *not* need to do — provide something everybody can reasonably afford — but is failing to do, with catastrophic effect, what it *does* need to do — regulate. The current degree of energy waste, pollution, carbon emission, and carnage, imposed by transport, is the visible result.

Aviation can emit, without charge, vastly more carbon than the same regional trip by rail. A train operator must compete with the car and regional bus services, yet is expected to offer a far greater degree of safety.

If the market is to provide transport, it must do so with a "level playing field" where every service is subject to the same taxation and regulation.

Perhaps governments can be forgiven for failing to recognize the importance of transport. Education and health are always priorities, in the eyes of any electorate. But fast, cheap, equitable, and above all adequate, transport is critical to any economy. It is not like any other industry, except post and telecommunications. These form *infrastructure*, on which every other activity depends. If their cost is excessive, everything suffers.

With the modal inversion from rail to road, transport costs rose, speed fell, and equality of access was lost. If road transport alone were adequate then congestion would not arise, incurring, as it does, a massive loss in production and curtailment of business. Economic growth may have been healthy since the war, in worlds both rich and poor, but its extent is surely limited by available capacity ('bandwidth') in all forms of communication, especially transport. Markets, new and old, must be reached and fed.

It is time to move forward again.

There are only so many reasons why we tolerate government and invest power in the State. First is the provision of security, without which, as we can see clearly abroad, there can be neither liberty nor prosperity.

Much is made of potential threats on the other side of the world. Vast amounts of our money are levied in taxes to pay for military adventures that aim to remove that insecurity. Much is also made of the threat of terrorism at home. Again, huge funds are raised and direct action is taken.

Even basic civil liberties are sacrificed to address the issue.

Meanwhile, innocent citizens are slaughtered by the thousand each year to preserve an almost absolute liberty in the use of the private car.

The State is failing to protect its citizens from far the greatest threat.

After security, we expect the State to promote prosperity by facilitating trade and industry. Currently, the most urgent issue with regard to the economy is overland transport, which is poor and failing.

Our mobility is actually lower than that of our grandfathers. For half a century now, Britain and America, in particular, have utterly failed to make use of our best technology, in this regard. A blind devotion to an obsolete mechanical contrivance has not only made real progress impossible, it has actually caused *regress*, and damaged both lifestyle and economy.

An entirely different lifestyle and economy may be waiting for us. A clean, safe, and healthy environment is certainly possible. An economy of a different order is also possible, were we to even begin to properly exploit the automation of which we are now capable.[24] Being inherently impossible to automate, road transport can have but little part to play. Since it is open to automation, rail is as important to our future as it was to our past.

It is vital to understand that the value of labour is fast diminishing, while the value of capital is rising. This process is inexorable. The second industrial revolution will prove as impossible, and as undesirable, to halt as the first. Only one phase of production could be automated last time; this time it is everything, even retail and distribution.

As before, we are faced with a simple choice: a new economy, or no economy. To survive, two changes must somehow occur.

24 It is perhaps no accident that the French, who have arguably embraced technology more than any other nation, can enjoy a shorter working week, and a higher average income.

For a large market to exist, the majority of the population must possess an income: the larger the income, and the larger the majority, the greater the market. In Britain, almost a third already depend on state 'benefits'. It seems doubtful that this proportion can rise any higher, since any increase in 'redistribution' requires either more taxpayers or higher taxation. In any case, any such income can only ever be marginal. With employment, and income from employment, set to fall, some alternative must be found.

There is only one other possibility: the ownership of capital must expand dramatically, into the population at large. No other self-sustainable model for any significantly automated economy is apparent. One mechanism by which this might progress is the reward of employees through shares, in addition to cash, as is already the practice in some companies.

The other vital sea change is in transport.

A modal inversion from rail to road has been seen by many to have been 'inevitable', and to simply have emerged from technological progress. Such is not the case. The very same progress could have revolutionized rail, as well as road. On both sides of the Atlantic, both governments and the doyens of industry did much to influence events, as we have seen. In any case, the catastrophe could easily have been averted, with modest intervention.

In Britain, the railway survived, despite mortal injury. Business is now booming, thanks to private investment and worsening road congestion. The line on which I live was once a main line. It is now to become one once again, and without a penny from Johnny taxpayer.[25]

In my home town of Oxford, the local newspaper now brims with calls for a clean electric light railway to replace polluting buses. A growing number of British cities are catching up with their continental cousins, which already enjoy comprehensive electric rail transit systems. The return of the bicycle, and arrival of the LEV, is also well underway.

Across Europe, a common system is arriving that will afford train automation.[26] Slowly, Britain is joining the high-speed rail network that now spans a continent, and which continues to grow.

---

25 The line is to be used from 2012 as a second express (160kph) route from Oxford to London. Plans are also advanced to reopen services to Milton Keynes and beyond.

26 The European Train Control System (ETCS), along with a common communication system, forms the European Railway Traffic Management System (ERTMS).

The situation for America is, of course, very different, though awareness of need is becoming evident. Despite the prevailing economic gloom, President Obama has offered the first US funding for high-speed rail.[27]

Only a *complete* reform of transport will afford the "bright new beautiful tomorrow" promised by Walt Disney, on his Carousel of Progress.[28]

When something that forms a basic part of every day life for almost everyone turns out to be harmful and antisocial, it can take a very long time for the truth to become accepted. Hostility and derision are to be expected.

If you had suggested to anyone a half-century ago that the day would dawn when smoking in a public place would become intolerable, even illegal, they might not have believed you. Smoking was not just considered social then; among males, at least, it was almost obligatory. It would be hard for any young man to be accepted by his peers if he did not partake.

Something deeply entrenched in our self-image, our society, and even in our everyday function, *can* change. New behaviour *can* be learned. Something cherished, even highly addictive, *can* be given up, if the reward is there.

Given the cost of the car, the reward is surely ample.

---

27 In January 2010, President Obama announced $8bn in grants for high-speed rail development, drawn from the $787bn economic stimulus package Congress approved in 2009, adding that "it creates jobs immediately and it lays the foundation for a vibrant economy in the future." There is only one (moderately) high-speed route currently in the US.

28 The Carousel of Progress was created by Disney himself and his WED Enterprises for the General Electric pavilion at the 1964–5 World's Fair in New York. It now resides in Tomorrowland at the Magic Kingdom, Walt Disney World. It portrays how the life of a 'typical' American family has been repeatedly transformed through technology.

Further Reading

Chapter 2 : *Ways and means*

Transport and land-use planning

1 Banister, David. 2002. *Transport planning.* Second Edition.
Taylor & Francis. ISBN: 0-415-26172-4.

A respected textbook on transport planning by a leader in the field. (David Banister is Professor of Transport Studies at The University of Oxford.)

Provides a "reflective analysis", looking back at transport over the last fifty years (predominantly in the UK), and a "comparative analysis" of experience in Europe and the US.

2 Mogridge, Martin. 1990.
Travel in towns : Jam yesterday, jam today and jam tomorrow?
Palgrave Macmillan. ISBN: 0-333-53204-X.

A seminal account of the relationship between speed and flow and the origin of congestion, among other things. His analysis is formal, and so is not for the faint hearted, but it represents science at its best and remains highly readable.

3 Downs, Anthony. 1992.
Stuck in traffic : Coping with peak-hour traffic congestion.
The Brookings Institution. ISBN: 0-815-71923-X.

The views of the American economist responsible for *Downs' Law* (see p.96). Discusses the US position, with so much car-predicated development.

Followed in 2003 by *Still stuck in traffic.*

4 Gehl, Jan. 1971. *Life between buildings : Using public space.* Arkitektens Forlag.
www.rudi.net/bookshelf/classics/lifebetweenbuildings/[1]

Jan Gehl has dedicated his life to improving the urban environment and is largely responsible for the one of the world's largest car-free public spaces — Strøget, in Copenhagen.

5 Transport for London. 2009. *Travel in London.*
www.tfl.gov.uk/corporate/about-tfl/publications/

Fascinating account by an effective transport authority showing how people use bus and rail in London at the Millennium. It suggests what can be achieved when authority over all modes is unified with an eye to co-operation (integration) rather than competition.

6 UK Department for Transport. 1996–2008. *National travel survey.*
www.dft.gov.uk/pgr/statistics/datatablespublications/personal/

A wealth of useful data based largely upon a sample of ~10,000 'representative' households. It's the best available data set for Britain, and is an official Government reference.

Also on the DfT website is *Transport Statistics* — an annual publication for the UK, which includes usage and casualty statistics for all modes, back to 1950.

Railways

7 Henshaw, David. 1991. *The great railway conspiracy.* Leading Edge. ISBN: 0-948135-48-4.

Records the decline of Britain's railway, and the corruption which aided it.

8 Booker, Frank. 1977. *The Great Western Railway.* David & Charles. ISBN: 0-715374-55-9.

A concise and readable history of "God's Wonderful Railway".

1 Availability and reference ("uniform resource locator" or URL) are subject to change. Each URL cited here can only be guaranteed correct at time of writing.

9 Jack, Ian. 2001. *The crash that stopped Britain.*
Granta. ISBN: 1-86207-468-2.

An account of the crash that exposed the true extent of neglect of railway infrastructure in Britain and the failure to instil effective safety management.

10 Atkins, for UK Department for Transport. 2004.
High speed line study : Summary report.
www.dft.gov.uk

Concludes a strong case for a new LGV running north-south in Britain.

11 UK Commission for Integrated Transport. 2004.
High speed rail : International comparisons.
www.cfit.gov.uk

Investigates why the only *Ligne de Grande Vitesse* thus far built in Britain cost several times more than its counterparts in continental Europe.

12 Murray, Iain. 2005.
No way to run a railway : Lessons from British rail privatization.
Adam Smith Institute, London.
www.adamsmith.org/publications/transport/no-way-to-run-a-railway/

Concise, though incomplete, summary of US experience.

Not surprisingly, given the author's institution and publisher, argues strongly in favour of a pure free market approach. An unquestioned assumption is that the playing field will thus become level.

Road and car

13 Plowden, William. 1971. *The motor car and politics in Great Britain.*
The Bodley Head, then in 1973 by Pelican. ISBN: 0-14-021628-6.

Plowden was a journalist, academic, and civil servant, eventually a member of the Central Policy Review Staff in the British Government Cabinet Office. He tracks the rise of the private automobile from the 1890s to 1970 and records the attempts made, too little and too late, to restrain it.

This history leaves no doubt of the abuse of privilege in the early days, or of the awareness in government of the problems posed by "unrestrained growth" in motoring.

A fascinating history from one associated with legislation through the explosive ascent to dominance of road transport over so many fields of policy.

14 UK Ministry of Transport, Colin Buchanan (Chairman). 1963.
Traffic in towns : A study of the long-term problems of traffic in urban areas. Her Majesty's Stationary Office (HMSO).

A highly influential report, ahead of its time in recognizing that new roads induce additional traffic (Downs' Law).

15 UK Ministry of Transport, R. J. Smeed (Chairman). 1964.
Road pricing : The economic and technical possibilities. HMSO.

An early argument for road-space charging to combat congestion. Technical difficulties in its implementation made it impracticable at the time.

16 Paterson, Matthew. 2007.
Automobile politics : Ecology and cultural political economy.
Cambridge University Press. ISBN: 0-52-169130-3.

A contemporary examination of how the car became so dominant, and a reflection on the political constraints on relieving its impact.

17 David Maddison *et al.* 1996.
Blueprint 5 : The true costs of road transport.
Earthscan. ISBN: 1-85383-268-5.

An academic summary of the cost of the car. Includes an attempt to assess road casualties in economic (monetary) terms. (Title refers to a series that began with *Blueprint for a green economy.*)

18 Paine, Chris. Video, 2006. *Who killed the electric car?* Sony Pictures.

The state of California once passed a law requiring car makers to supply a proportion of new vehicles with zero emission. Their response was to deploy battery technology. Opposition, and slow sales, eventually defeated both the technology and the law. The story is a telling one, with regard to vested interests and to the market for private vehicles.

Chapter 3 : *A few short stories*

19 UK Driver and Vehicle Licensing Agency(DVLA). 2004.
For medical practitioners : At a glance guide to the current medical standards of fitness to drive. HMSO.

How a doctor or paramedic should respond *if a patient seeks their advice*, including whether they are then obliged to inform the DVLA, with the possibility of licence withdrawal.

Chapter 4 : *Sustenance*

Energy

20 Schumacher, Ernst Friedrich. 1973.
Small is beautiful : A study of economics as if people mattered.
Vintage (Random House). ISBN: 0-09-922561-1.

A seminal work by a brilliant individual.

Very briefly, he introduces two ideas into economic analysis: first, that some resources are finite and that the cost of their depletion should be counted; second, that people matter, and so the cost of factors such as employment and impacts upon health should be counted also.

Schumacher predicted many things that have come, or are coming, to pass, including conflict over diminishing energy resources, and that nuclear power installations would leak. He was deeply critical of nuclear power, commenting that no degree of leakage was acceptable.

21 Scheer, Hermann. 1999.
The solar economy : Renewable energy for a sustainable global future.
Earthscan. ISBN: 0-85383-835-7.

Advocates a move to an economy based on solar energy.

Compares the "supply chain" of fossil fuel and solar energy, and discusses the economics of both (with a brief mention of nuclear).

22 World Energy Council. 1994. *New renewable energy resources.* Kogan Page. ISBN: 0-7494-1263-1.

A study of the potential for a variety of renewable energy sources.

23 Appenzeller, Tim. 2004. *The end of cheap oil.* National Geographic, **205**, 6 (June), pp. 80–109.

An explanation of why oil will become progressively more expensive.

24 Parfit, Michael. 2005. *Powering the future.* National Geographic, **208**, 2 (August), pp. 2–31.

A summary of our energy options, as fossil fuel becomes less affordable.

25 Bourne, Joel K. jr. 2007. *Biofuels : Boon to boondoggle.* National Geographic, **212**, 4 (October), pp. 38–59.

An assessment of biofuels with regard to price and environment.

Global warming and its impact

26 Houghton, John. 1997. *Global warming : The complete briefing.* Second Edition. Cambridge University Press. ISBN: 0-521-62932-2.

A thorough survey of the subject by someone with the highest academic and professional credentials.

Sir John Houghton CBE FRS was for many years Head of Atmospheric Physics at Oxford University and Chief Executive of the UK Meteorological Office. He then went on to chair the UK Royal Commission on Environmental Pollution and co-chair the Scientific Advisory Group of the Intergovernmental Panel on Climate Change (IPCC).

27 Stern, Sir Nicholas. 2006. *The economics of climate change.* HMSO. ISBN: 0-521-70080-9.

Highly influential review which cast the costs of climate change in economic terms. As well as putting a figure to the current external cost incurred by emitting each tonne of carbon dioxide, Stern also spelled out the economic opportunity offered by the market for new applicable technology.

28 Appenzeller, Tim et al. 2004. *The heat is on.*
National Geographic, **206**, 3 (September), pp. 2–75.

A review of the evidence for global warming, the contribution made by human activities, and possible consequences for planetary climate and ecology.

29 Hayden, Thomas. 2006. *Super storms.*
National Geographic, **210**, 2 (August), pp. 66–77.

A summary of how hurricanes form and the likely effect of rising sea-surface temperature.

30 Appenzeller, Tim. 2007. *Big thaw.*
National Geographic, **211**, 6 (June), pp. 56–71.

The state of glaciers and ice sheets, and the consequences of their melting.

Transport and carbon emission

31 Inderwildi, Oliver, *et al.*, Prof. Sir David King FRS (Ed.). 2009.
Future of mobility road map.
Smith School of Enterprise and Environment, University of Oxford.
www.smithschool.ox.ac.uk/wp-content/uploads/2010/02/Future_of_Mobility.pdf

An authoritative review of the status of transport with regard to carbon emission. Many conclusions parallel those in this book, such as the inability of biofuel to substitute for fossil fuel due to the limited availability of productive land, and the dominant rôle that rail has to play in removing our extreme reliance upon the car (Figure 7.2, p. 89).

32 Sperling, Daniel, and James S. Cannon. 2006.
Driving climate change : Cutting carbon from transport.
Academic Press. ISBN: 0-12-369495-7.

After a summary of the current situation, two approaches to reducing carbon emission from transport are discussed: promoting a shift to modes that consume less energy, and the development and deployment of new energy technology.

Pollution

33 Royal Commission on Environmental Pollution. 1997. *Twentieth report : Transport and the environment — developments since 1994.* HMSO. ISBN: 0-10-137522-0.

www.rcep.org.uk/reports/

Where scientific authority stepped in, and change began in Britain.

Many other reports of interest can be found at RCEP website.

Chapter 5 : *Routes and urbs*

Urban planning and design

34 Howard, Ebenezer. 1902. *Garden cities of tomorrow.* MIT Press. ISBN: 0-262-58002-0.

A landmark publication that founded the "garden city" movement in post-war Britain. Unfortunately, Howard could not have foreseen the advent of mass car ownership, or how it would affect the evolution of both town and country, which he sought to unite.

35 Le Corbusier. 1929. *The city of tomorrow and its planning.* Dover. ISBN: 0-486-25332-5.

The champion of the 'modernist' movement, Le Corbusier advocated complete accommodation of the car, relegating humans to huge tower blocks.

36 Hall, Peter. 1988. *Cities of tomorrow : An intellectual history of urban planning and design in the twentieth century.* Basil Blackwell. ISBN: 0-631-17567-9.

Contains an authoritative discussion of both Howard and Le Corbusier, and their influence on urban planning throughout the world.

Walt Disney and EPCOT

37 Mannheim, Steve. 2002. *Walt Disney and the quest for community.* Ashgate Publishing. ISBN: 0-7546-1974-5.

An account of Walt Disney's concern over the "urban crisis" that emerged in the 1960s, with the violent breakdown of social cohesion and consensus in America, and his response — the *Experimental Prototype Community of Tomorrow* (EPCOT).

38 Disney, Walt. Video, 1967. *EPCOT.* Walt Disney Productions. In *Walt Disney Treasures : Tomorrowland.* 2004. Walt Disney Video.

Disney recorded his intentions just a couple of months before he died.

Made as a presentation to legislators, investors, and the media, it is thoroughly captivating. Whether or not you are convinced, you can only sit back in awe at the boldness of the venture, and of the man behind it.

39 Thomas, Bob. 1976. *Walt Disney : An American original.* Simon & Schuster. ISBN: 0-671-22332-1.

The classic biography of Walt Disney, written by someone who knew him.

Thomas worked alongside Disney, and some of his senior animators, to produce a documentary — *The art of animation.*

Planning for change

40 Banister, David. 2005. *Unsustainable transport : The transport crisis.* Taylor & Francis. ISBN: 0-415-35790-X.

A noted authority presents a coherent argument that current urban transport and land-use patterns are unsustainable.

After establishing the connection between land use and transport, he explores the obstacles to, and constraints upon, change, together with opportunities to exploit new technology. A way forward is finally proposed.

41 Hickman, Robin, Catherine Seaborn, Peter Headicar, David Banister, and Corinne Swain. 2010.
Spatial planning for sustainable transport?
Town and Country Planning, **79**, 2 (February), pp. 71–77.

An authoritative and readable article summarizing policy options for regional and urban development that minimize car dependence.

Concerned mainly with efficiency and congestion, it does not extend to social concerns, nor does it suggest any form of accountability. Some might thus object that its recommendations tend to dictate lifestyle. Nonetheless, it demonstrates recognition of both the urgent need for change and the essential unity in transport and land-use planning.

Index